Technical Documentation

Technical Documentation

A. J. Marlow

# Technical Documentation

A. J. Marlow

Copyright © Andrew J. Marlow 1995

The right of Andrew J. Marlow to be identified as author of this work has been asserted in accordance with the Copyright, Designs and Patents Act 1988.

First published 1995

NCC Blackwell
108 Cowley Road, Oxford OX4 1JF, UK

Blackwell Publishers Inc.
238 Main Street, Cambridge MA 02142, USA

All rights reserved. Except for the quotation of short passages for the purposes of criticism and review, no part of this publication may be reproduced, stored in a retrieval system, or transmitted, in any form or by any means, electronic, mechanical, photocopying, recording or otherwise, without the prior permission of the publisher.

Except in the United States of America, this book is sold subject to the conditions that it shall not, by way of trade or otherwise, be lent, resold, hired out, or otherwise circulated without the publisher's prior consent in any form of binding or cover other than that in which it is published and without a similar condition including this condition being imposed on the subsequent purchaser.

British Library Cataloguing in Publication Data

A CIP catalogue record for this book is available from the British Library.

Library of Congress Cataloging-in-Publication Data
Marlow, A. J. (Andrew J.)
    Technical documentation / A. J. Marlow.
      p. cm.
    Includes index
    ISBN 1-85544-216-1 (pbk. : alk.paper)
    1. Technical writing. 2. Technology—Documentation—Data processing. 3. Technical manuals. 1. Title.
T11.M348   1995
808'.0666—dc20   95-12432
                                                      CIP

ISBN 1-85554-216-1

Typeset in 10 on 12 pt Times New Roman
Printed in Great Britain by Hartnolls Ltd., Bodmin, Cornwell
This book is printed on acid-free paper

Contents

Preface xi

Acknowledgements xiii

1 **THE AUTHOR** 1
    SOFTWARE TECHNICAL AUTHORS 2
    HARDWARE TECHNICAL AUTHORS 4
    THE AUTHOR'S ROLE 5
    SUBJECT KNOWLEDGE 7
        Writing Sources 7
    READER KNOWLEDGE 9
    COMMUNICATION SKILLS 12
        A Command of English and Grammar 12
        Communicating with Others 13
    ORGANISATIONAL SKILLS 14
    SUMMARY 15
    FURTHER READING 16

2 **TECHNICAL WRITING** 17
    PRELIMINARY PLANNING 17

vi TECHNICAL DOCUMENTATION

FORMATTING	21
LANGUAGE AND STYLE	23
EDITING	26
PROOF-READING	27
SPECIFICATIONS	28
HOUSE STYLES	36
Page Style	36
Headings	36
Spelling and Hyphenation	36
Punctuation	36
Capitals	37
Text Highlights	37
Abbreviations and Units	37
Dates	37
Notes and Reference System	37
Tables, Lists, Figures, Maps, Plates, etc.	37
Copyright and Trademarks	38
Indexing	38
PREPRESS PREPARATION	38
SUMMARY	39
FURTHER READING	40

3 TECHNICAL PUBLICATIONS 41

USER GUIDES	42
REFERENCE MANUALS	44
SUPPORT DOCUMENTATION	44
OVERHAUL AND SERVICE DOCUMENTATION	45
TROUBLESHOOTING GUIDES	46
REPORTS	46
TRAINING MATERIAL	46
UPGRADES AND REVISIONS	47
SUMMARY	48
FURTHER READING	49

4	**DOCUMENTATION DESIGN**	51
	DESIGN AND LAYOUT	52
	Page Size	52
	Page Layout	53
	Text Columns	53
	Other Page Elements	55
	TYPOGRAPHY	59
	Typefaces	59
	Typesize	60
	Typestyles	60
	Spacing	62
	COLOUR	62
	FINISHING AND PACKAGING	63
	Paper Sizes	64
	Trimming and Binding	66
	Ring Binding	68
	Mechanical Binding	68
	Perfect Binding	69
	Sewn Binding	69
	Case Binding	69
	Imposition	69
	Types of paper to use	71
	SUMMARY	72
	FURTHER READING	73
5	**WORD PROCESSING**	75
	WORD PROCESSING FEATURES	75
	Editing Functions	76
	Spacing, Alignment and Other Formatting	77
	GRAMMAR AND STYLE CHECKERS	78
	Spelling Checkers	78
	Thesauruses	79
	Grammar and Style Checkers	79

viii TECHNICAL DOCUMENTATION

AUTOMATIC REFERENCING AND INDEXING	84
SUMMARY	85

6 ELECTRONIC PUBLISHING — 87

DESKTOP PUBLISHING	87
Page Make-up and DTP	89
Style Control	89
Graphics	91
Output Devices	92
Methods Employed	92
CATEGORIES OF DTP SYSTEMS	92
Group I — Word processor-based systems	92
Group II — PC-based page make-up systems	92
Group III — PC-based electronic publishing	92
Group IV — Dedicated electronic publishing systems	92
WORD PROCESSOR-BASED DTP	93
PC-BASED PAGE MAKE-UP	94
ELECTRONIC PUBLISHING	96
DEDICATED SYSTEMS	97
DATABASE PUBLISHING	98
SUMMARY	99
FURTHER READING	99

7 GRAPHICS AND ILLUSTRATION — 101

GRAPHICS AND DRAWING SOFTWARE	102
Image File Formats	104
BMP/DIB/RLE File Formats	104
GIF File Formats	105
PCX File Formats	105
TIFF File Formats	105
IMG File Formats	106
MAC File Formats	106
TGA File Formats	106
Vector File Formats	106

	INCLUDING GRAPHICS IN DOCUMENTS	106
	POSITIONING OF ILLUSTRATIONS	108
	REFERENCING ILLUSTRATIONS	109
	CAPTIONS	110
	SUMMARY	110
	FURTHER READING	111
8	**ON-LINE DOCUMENTATION**	113
	HELP SYSTEMS	113
	HYPERTEXT AND MULTIMEDIA SYSTEMS	114
	SUMMARY	116
9	**PUBLICATIONS MANAGEMENT**	117
	THE PUBLICATIONS MANAGER	117
	PLANNING DOCUMENTATION PRODUCTION	118
	RECRUITING AND AUTHOR MANAGEMENT	119
	PRINT BUYING	120
	Print Specifications	121
	Supplier Selection	123
	IN-HOUSE PUBLICATION SYSTEMS	124
	COSTING AND SCHEDULING	125
	Costing	125
	Scheduling	129
	QUALITY CONTROL	131
	SUBCONTRACTING	133
	DOCUMENTATION DEVELOPMENT	136
	REVISION MANAGEMENT	137
	SUMMARY	139
10	**INTERNATIONAL DOCUMENTATION**	141
	WRITING FOR INTERNATIONAL AUDIENCES	141
	TRANSLATION	143
	SUMMARY	143
	FURTHER READING	144

TECHNICAL DOCUMENTATION

11	**TRAINING AND EDUCATION**	145
	AUTHOR TRAINING	145
	NATIONAL QUALIFICATIONS	145
	City and Guilds	145
	GLOSCAT Diploma in Software Documentation	146
	Higher Education	147
	INDEPENDENT COURSES	148
	DISTANCE (OPEN) LEARNING	152
12	**LIST OF STANDARDS**	153
	STANDARDS	153
13	**ORGANISATIONS AND INSTITUTIONS**	157
	THE INSTITUTE OF SCIENTIFIC AND TECHNICAL COMMUNICATORS	157
	THE SOCIETY OF AUTHORS AND THE SCIENTIFIC AND TECHNICAL AUTHORS GROUP	160
	SOCIETY OF FREELANCE EDITORS AND PROOFREADERS	161
	SOCIETY OF INDEXERS	162
	TECHNICAL DOCUMENTATION SERVICES	162
14	**GLOSSARY**	171
	INDEX	183

Preface

The subject of technical documentation is so great, that this one volume cannot do it justice. However, its aim is to provide a comprehensive outline to documentation origination and production management. It is also intended to provide a key source of background information for those studying for examinations or practising authorship or publications management in all technical fields, but with certain emphasis on the computer industry.

The book will serve as a valuable aid to those students preparing for examinations and individuals intending to undertake technical authorship or technical publications management as a career.

The occupational role of a technical author is under constant development. Since the skills are so often accompanied by the use of technology to achieve results, the speed at which the technology changes influences the rate at which the author/manager needs to acquire new skills. This book describes some of the technology used in the publication of technical documentation, though it is subject to considerable change in rather short timescales, so no specific detail is given of particular products or systems.

The future of the occupation is discussed in so far as it is generally recognised that the skills of today's technical authors may be put to a future use which does not involve the production of printed matter at all. To this end, there are some who would like to see the title 'technical author (or writer)' replaced with a more generic label, such as 'technical communicator' or maybe 'information design consultant'. Despite intentions to introduce new job titles, the uninitiated is just as bewildered about what a technical author or technical publications manager actually does and the breadth of skills and knowledge required to do the job will become clear in the forthcoming chapters.

Throughout this book, I have chosen to distinguish between two particular areas of technical documentation as an occupation; technical authorship and publications management. (The term 'author' will be used rather than 'writer', though I do so through no particular preference.) While some individuals may find themselves tending towards one or other of these two occupations, there are many employed in the production of technical documentation who both write and manage the publication of technical manuals, guides and other literature.

Apart from providing useful material for students and budding technical authors/ managers, it is hoped that this book will be a useful reference tool for anyone whose job responsibilities extend to the origination and publication of technical documentation.

As briefly mentioned previously, this book tends to place emphasis on technical documentation in the computer industry. There are two reasons for this. Firstly, it is the industry in which I have had most experience in this particular field. Secondly, it is one of the fastest-growing industries within which the role of the technical author and publications manager is becoming increasingly important. Most technical and engineering developments are related in some way to computer technology. Most of the subject matter covered in this book is as relevant to the engineering technical author as it is to the software author. Although I have made some distinctions between the two in the chapters concerning the role of the author, I personally believe that there is little difference between the skills needed to do the job in either case if one ignores specialist product knowledge, though I know of some technical documentation specialists in the engineering industry who would disagree strongly. I hope that in the rest of this book, I have been as objective as possible and indicated where subjective, personal options are expressed.

Dr Andrew Marlow BSc MSc DBA

Titchmarsh

Acknowledgements

The author wishes to thank the many organisations who have provided information about their services, to the Institute of Scientific and Technical Communicators and to the British Standards Institution. Extracts from British Standards are reproduced with the permission of BSI. Complete copies can be obtained by post from BSI Customer Services, 389 Chiswick High Road, London, W4 4AL; telephone 0181 996 7000.

1 The Author

This chapter describes the role and skills of the technical author. Ask a number of technical authors or publications managers for the definition of a technical author and you will get a range of conflicting answers. This is partly due to the fact there are few specific job descriptions for technical authors, either in education or industry, and very few broadly recognised qualifications that relate to the career. Technical authors cannot even agree on their title among themselves; some prefer to be called technical writers, or technical communicators, or information managers, among other choices. The reason why the term technical author is not readily accepted is because individual authors view their responsibilities differently. Some feel that 'author' or 'writer' is too limiting and implies their jobs are similar to a novelist, except that they are producing words on technical subjects. Indeed, whatever title one chooses, there is more to being a technical author than writing.

The view of the technical author has changed dramatically over just the past few years. Such a previously obscure job has become a recognised profession of equal standing to other qualified individuals, such as engineers or programmer/analysts. This may not be a widely held view, but recognition of the status of technical authors is growing, along with the importance of quality documentation. However, as a result of conversations with many individuals in the profession whom I have had the opportunity and pleasure of meeting, I have concluded that the position and status of a technical author or a publications manager varies noticeably from one organisation to another.

For some authors, the difficulty still remains of securing for themselves, and their work, the same degree of respect and commitment as that received by other departments within their organisations. In this situation lies the source of many of their problems regarding the degree of success or failure in being able to make a significant enough contribution to producing good quality documentation. Few senior members of the management in organisations are familiar with the role of the technical author and most have no or little experience in this field of work. Where the author is not part of a publications team, the author reports to either a technical department superior or to a sales and/or marketing executive.

It is partly the way in which the authors are regarded by others in their own organisation that affects their ability to produce high-quality technical documents. If the author is viewed as simply a technically minded individual who writes 'the manual', then this is no more a comprehensive understanding of an author's role as suggesting that the role of a sales representative is simply that of selling a product or service. Such is a very limited description of the responsibilities of a sales representative that include maintaining the loyalty of customers; providing appropriate reassurance in the event of problems; acting as a consultant and advisor on behalf of the company; being responsible for forecasts and budgets; increasing the sales potential to get more job satisfaction and the improvement of company profits.

All of the above qualities are easily recognisable in the commercial sense. They seem to be appropriate to the role of a sales representative. Yet, perhaps surprisingly, many of them are also areas of responsibilities of successful technical authors and publications managers, whose work carries the organisation's own image and reputation to the reader. Few would include such characteristics in a list of author responsibilities, especially those who select and employ authors in the first place.

Technical authors have a responsibility to communicate and inform. They are often the bridge between expert technical knowledge and the layperson. Their work is rewarding in the sense that the job produces a tangible result. Their work is also open to immediate criticism, since the user of technical documentation will not tolerate errors, omissions and ambiguities. Once such problems are encountered, the reader loses faith in the documentation and the author immediately. The author must be well trained and needs to be equipped with the skills appropriate to his or her area of specialisation. The following paragraphs discuss the two main fields of technical authorship, role of technical authors and the key elements of the job.

SOFTWARE TECHNICAL AUTHORS

Computer software is one area in which technical authorship has developed rapidly. This is mainly due the considerable range of software products in the world-wide market that are supported by documentation. As a result of the common application of computers and software in homes and businesses, many more people have had cause to use technical documentation; there are far more readers of technical documentation than ever before. Prior to the boom in computer software on personal computers, most ordinary folk only encountered technical documentation in the form of instructions that accompanied a domestic appliance or their family car. Other technical documentation of any extent was only ever encountered in the course of specialist study and work. Now, with computer software so commonplace and affecting all our lives, people are coming into contact with far more documentation, some of it of considerable extent, and using it as support for their business software, computer games and other leisure applications.

The development of desktop publishing — itself an application of software — means that the tools to produce documentation are available to more individuals and the whole process of producing and publishing documentation has become much easier. The result is yet more user guides and manuals. Like television, the computer is a medium for communication in its own right. The screen conveys information to the user and this means that the scope for the software technical author extends far beyond the printed word. The skills and experience of software technical authors often encompasses the application of multimedia communication devices; the use of audio and video material.

Since the computer screen can present information to support the software in use at the same time (in the form of messages, prompts, help screens and graphics), the objective of the printed documentation changes. There are many software user manuals that do little more than repeat the information on the computer screen, albeit presented in a different order. Their effectiveness is questionable. A well-designed software application that is intuitive to use should, in effect, need little documentation to support it. In view of this, the technical author working on a manual to accompany a software application needs to give serious consideration to objectives and purpose of the documentation. Since the variety of software applications is so great, the software technical author is faced with as many different areas of specialisation as authors in other fields of technology.

The software technical author may be involved in the production of on-line documentation; that is, information available to the user on the computer screen, perhaps in the form of indexed or context-sensitive help text and has far more devices available for communication than an author writing a user guide to accompany, say, a hi-fi system or microwave oven.

One aspect of software manuals that distinguish them from other technical documentation, is the marketing element. The user manuals that accompany a packaged software application for a personal computer are often the most significant component of the package. Software manuals are far more 'market conscious' than other technical documentation such as that produced in the fields of science and engineering. There is a certain value that is added to a software product that includes extensive, glossy manuals. Without these, the user would receive only one or two diskettes or CDs and that may not be perceived as value for money. Consequently, software technical authors often find themselves producing documentation than is more than merely informative. It needs to support the marketing of the product and convey other underlying messages, such as corporate image. The finishing and packaging of software documentation carries far more significance than other technical documentation. The software technical author therefore needs to be 'market aware'.

While writers of documents supporting application software rarely have the burden of the user's health and safety to consider, except in certain specialist areas, such as medicine, poor documentation can lead to expensive and inconvenient consequences for users. The software technical author must therefore be as vigilant in the pursuit of accuracy and completeness as any other, and because the potential readership can be huge (consider how many users there might be for Microsoft's MS-DOS user guide), the repercussions of poor documentation can be equally great.

Software authors need to work closely with those developing the software products about which they write. Unfortunately, due to the highly competitive, commercial outlook in the software industry and the huge range of products that can be developed by individuals working on their own, standards and approaches to technical writing vary greatly. There are few standards to follow, with the exception of the British Standard BS7649 — *The design and preparation of documentation for users of application software*, and this is unlikely to have much impact on the quality of software documentation in the short term. Unlike hardware products, many software products are designed 'on-the-hop'. There may be no blueprint to work from or even an outline specification and the documentation is often left to the last minute, once the software product has been developed. Software authors must therefore learn to work in parallel with the development activities and to encourage this relationship where it does not already exist.

Software documentation projects have a quite different development cycle to hardware and other engineering-related systems. Where a manufacturing process is involved, the time between conception of the product and final delivery from the works can be much greater than that of software, giving the author more time to prepare the documentation. Because of the 'fluid' nature of software design, a software product may be developed and enhanced right up to the day of despatch, and the common tendency is for software developers to do so. This often results in documentation projects having to be completed in advance of the product in order to satisfy print and production timescales, but without prior knowledge of the impending changes or enhancements. Software documentation is frequently out of the step with the product on the day of release. The software author must be adaptable to such circumstances and to have the skills to manage a project in these cases. Precise scheduling of the documentation project can help, but the liaison between author and developer is perhaps the key requisite for successful software technical authorship when timescales are short.

HARDWARE TECHNICAL AUTHORS

Technical writing for hardware covers a much wider field than software technical writing. It encompasses engineering, electronics, aviation, plant and equipment among others. Health and safety aspects in this field of authorship are far more prominent. In many situations, inaccurate or misleading writing can lead to injury or some other disaster, since much of the documentation produced is designed to ensure the safe and efficient operation of equipment. Technical documents relating to hardware are usually required at more levels than software manuals. In the case of software documentation, almost all of it is aimed at the user's level. For hardware, the user's guide may be the least significant piece of technical documentation. Technical specifications, service, repair and maintenance manuals may be far more extensive and will be aimed at quite different levels of readership.

The responsibilities carried by the hardware technical author are often greater than that of the software author. A software author may be able to produce a reasonable user guide without specialist technical knowledge and with little more experience than the end user. For hardware technical authors, this is rarely the case. One finds probably the deepest level of technical detail and engineering procedures in a repair and overhaul manual for a complex piece of equipment. It may include detailed technical descriptions, test procedures, dismantling and reassembly instructions, etc. Although in such circumstances, the author is writing for highly skilled personnel rather than the layperson, no less attention should be given to clarity and ease of use in the documentation.

Accuracy is, without doubt, extremely important. The success of special areas of hardware documentation, such as fault diagnostic information, relies heavily on sound understanding of the underlying principles. The hardware technical author will be better equipped if he or she has practical experience in the technology, or at least has the ability to question experienced service personnel on, say, relating symptoms to probable causes.

The consequence of the differences between software and hardware technical authors is that, in general, hardware technical authors tend to be more qualified in their field of operation. This does not mean, however, that they are any more qualified in technical authorship. Training in this area is as important to the hardware technical author as it is to any other technical communicator.

THE AUTHOR'S ROLE

The main role of a technical author is to ensure the effective communication of technical information in whatever form appropriate. The result of the authors work should be to have provided information that is complete, accurate and easy to understand by those who will use it. In order to achieve this, the author must be skilled. Adequate training is essential, but often missing. Many individuals who happen upon technical authorship often do so as a 'second career' move, usually having been employed in a technical capacity relating to the subject.

Many companies do not recognise the need for technical author's and rely on programmers or application engineers to write the content with the graphics and publishing functions carried out professionally. It is often regarded that the author's job is simply to produce 'words'. However, in order to carry out their role effectively, the author must successfully employ a number of related skills as are listed on the following page.

As implied earlier in this chapter, the role of the author can vary widely according to responsibilities assigned by the organisation in which he or she works. On the one hand, authors may work in a team as part of a complete publications unit or department, on the other, they may work on their own and have responsibilities that encompass all aspects of technical publication. The role must defined within the local needs of the job. Above all, the author's role involves communicating technical information. In one sense the author is a technical researcher, since the acquisition of information, its analysis and use form the largest part of the author's workload. The actual writing of the documentation should be a routine part of the job, while the effective employment of all other skills makes the difference between a good and poor quality author.

There are many elements that define the role of the author in the production of technical documentation. The author must be able to:

(a) determine the purpose of the document;

(b) target the audience and analyse their needs;

(c) develop a plan for a document that will achieve the objective set in (a);

(d) collect, analyse and evaluate the data necessary to compile the document;

(e) choose an appropriate style and form of presentation for the information;

(f) organise the document logically and consistently;

(g) write clearly and unambiguously;

(h) design an appropriate method of production, taking into account the way in which the document will be used and subsequently updated;

(i) find and engage appropriate resources for evaluation and quality control;

(j) find and engage appropriate resource for production;

(k) control the on-going maintenance of the documentation;

(l) employ means to improve quality, cost-effectiveness and efficiency in the documentation cycle.

As with any other type of work that involves a production process, designing, writing and printing a technical document often has to be done within commercial constraints. This is especially true of subcontracting authors, whose productivity is more obviously connected with economical factors.

Despite the fact that all authors should strive to produce documentation of the highest standard, two limiting factors will force compromise: time and cost. In the words of a company director to a technical author: 'I don't want it good, I want it Monday'. Attention to detail is still important, however, even with such constraints. It is certainly one of the least glamourous aspects of technical writing that, for example, indexes are painstakingly compiled, or that the sentence structure is examined more than once to ensure that the information is presented in the best possible light. The author must be able to concentrate over significant periods of time to be able to achieve this.

SUBJECT KNOWLEDGE

It is generally accepted that writers of technical documentation should have at least some knowledge of the subject about which they will be writing. The degree of knowledge required will depend upon the amount of input to documentation that may be provided by others. There are those who believe that authors must know, in great depth, the subject about which they are writing if the documentation is to be at all meaningful and comprehensive. However, good documentation can be provided by an author who has only a general grasp of the topic, yet who is able to assimilate accurate information from appropriate sources where the subject knowledge abounds, and present it in an informative way.

Having a complete understanding of the subject is clearly to the author's advantage, but care must be taken to ensure that the author does not lose sight of the level of knowledge expected by the reader. It is all too easy to gloss over familiar concepts or inadequately explain something that is obvious only to the author. In circumstances where the author of a technical document is also an expert in the subject, it will help if the work is read by someone whose own level of knowledge is similar to that expected of the intended readership; particularly where the documentation is intended to inform laypersons.

If the author is relying on others to provide the technical information for the subject of a document, it is important that the work is vetted by the experts concerned to ensure that the author has not misrepresented or misinterpreted the information which may cause confusion to the reader. Where an author is unsure of the subject matter, no attempt should be made to extrapolate technical information supplied. This, too, will result in either ambiguity or falsehoods and will serve no use to the reader. Checking the accuracy of technical content is vital where the subject knowledge of the author is limited and the author must be vigilant in the pursuit of clear facts on which to base any written material. Similarly, technical diagrams and other illustrations must be precise and checked by those with suitable authority.

If the author is given the task of writing documentation without the input or aid of others who have the technical know-how, the author must be able to adapt to new projects and cope with short learning timescales. A thorough understanding of the requirements of the documentation will also be required; it may not be obvious when dealing with an alien subject. Only then can the author be expected to present the information precisely and clearly enough.

Writing Sources

Not all documentation is, by any means, written by one individual. Input to technical documentation within industry comes from a wide variety of sources. The technical author may be only one of the many resources involved in the writing and input may be so significant from other parties, that the job of the author can become more like that of an editor.

Subject knowledge is just as important in these circumstances, since one cannot edit effectively without the authority of that knowledge. When producing user documentation for a technical product, for example, the technical or development departments may provide a significant amount of specification material which constitutes the groundwork of the user documentation itself. In the software industry, the availability of a well-written specification for a product can certainly cut down the amount of time that the author has to invest in finding out the how the product works.

In the production cycle of a technical document, it may be that a working team is created to cover various aspects of documentation, each member of which may have particular skills or specialise in specific areas of subject knowledge. Such individuals may include training staff, whose job implies that they must understand the subject concerned, and may well provide significant input to the documentation. Indeed, some authors combine writing and training functions — both of which require optimum communication skills — and they represent some of the best technical authors.

In a national survey of documentation for the computer industry, organised by Digitext, an analysis was based on a question which asked for the percentage of involvement for each resource used to write documentation of varying types of project. The responses to the question were treated as instances of a writer's involvement, expressed as a decimal relating to the percentage (for example, 1 for 100%, 0.5 for 50% and so on). Table 1.1 shows the result of the analysis.

Digitext clarified the 'Other' category included engineering, support staff, marketing department and educational services, while the 'Other documents' category incorporated brochures, reports, fact cards and data sheets, services manuals, system specifications, quick references guides and so on.

Table 1.1 Percentage involvement for each resource in preparing documentation

	Technical reference manuals	Installation manuals	Training manuals	User manuals	Introduction manuals	Other documents	Total
Product developer	5.40	4.85	2.00	7.20	1.60	1.05	22.10
Project manager	1.40	.10	.10	2.10	.70	.00	4.40
Trainer	.20	.20	11.45	.60	.40	.00	12.85
In-house writer	10.30	7.25	9.80	14.05	8.70	5.35	55.45
Contract writer	.20	1.10	.00	1.45	1.60	.00	4.35
Documentation house	.90	.50	.15	4.30	2.00	2.60	10.45
Other	2.60	2.50	.50	1.40	1.90	3.00	11.90
Total	21.00	16.50	24.00	31.10	16.90	12.00	121.50

The analysis showed clearly the split of subject knowledge and how much involvement belonged to the technical authors. In conjunction with this, the survey also analysed, in a similar way, those responsible for reviewing the draft documentation with the development staff playing a more significant part.

Where subject knowledge is lacking in the author for a particular documentation project, it is important to consider alternative methods of documentation production.

One solution may be to employ a subcontracting author or documentation house which has individuals with the required skills and background knowledge.

READER KNOWLEDGE

It is unlikely that the documentation for a technical subject can be either interesting and/or useful to the reader if the author has failed to understand the reader's requirements. Knowing your reader is one of the most fundamental elements to ensuring the success of any technical documentation. The author should not make assumptions about the level of understanding of the reader. It is a matter easily overlooked. Many examples of technical documentation show that writers are good at describing what a piece of equipment or software does, but few offer constructive guidance on applying the technology described to particular objectives.

By way of an absurd but poignant analogy, imagine that you have to write an instruction manual on how to use a car and the reader has never used or even seen a car before. There is no point simply describing the technical features of the car, such as performance, specification, design, etc. The reader will not be able to interpret what to do with the car from this. Even if you outline the purpose of the car, describe the controls and explain what to do to go from A to B, you could easily overlook the most fundamental steps in the process. This is because of your familiarity with a subject that is second nature to almost everyone. For example, would you think to describe how to open the door to get in the car? Would you remember to explain that fuel must be put in the car before it can work and offer guidance on where to obtain it?

Operator manuals for cars do not begin with such fundamental details because the intended readership is likely to be well aware of them. It is far more difficult to draw the line between reader and author knowledge with less familiar subjects. There are ample examples of user manuals that, although comprehensive is most aspects, omit quite simple yet important information. They fail to explain some basic setup procedure (the 'how to get in the car' bit) or wrongly assume prior knowledge of some aspect of a subject. Many failed technical documents have proved unsuccessful simply due to a lack of understanding by the author of the reader's own level of knowledge.

Using unfamiliar terminology is a common problem which is associated with reader knowledge. Technical documentation should always provide an explanation of terms that are used and not part of everyday vocabulary. This is usually achieved by including a glossary of terms, but jargon is still commonly used which is often omitted from glossaries.

Documents that need to address a whole range of possible readership levels are best aimed at the lowest level unless different sections are clearly applicable to different levels of reader skills. However, it is important not to treat all documentation as if it must be understandable by the layperson. This is neither necessary or practical. If every manual addressed the readership level of a layperson, some would be extremely large indeed. There is nothing wrong in writing on the basis that certain prerequisite knowledge is required, provided one is sure that knowledge exists in the reader and that it not assumed without good cause.

Once the level of readership knowledge is established, it is important that the text is consistent in the way it addresses the reader. The author should avoid treating one paragraph as if read by a layperson, then suddenly assuming expert knowledge in the next. This mistake is often made inadvertently by the author who finds some topics easier to put across in layperson's terms than others. When reading such material, one can sense when the author has retreated to the safety of their own level of knowledge and has given up on the reader.

For some types of documentation, it may be relevant to investigate the reader's needs directly from potential readers. Some of this information may be available in the form of feedback as a result of the publication of similar technical documents. Otherwise, research will be required. Few authors devote any significant amount of time to analysing the readers of their work. Many work on assumptions. Even those authors who may be prepared to carry out some analysis of their readers and have the time to do so, may be unsure how to go about the task. A good starting point is to summarise which aspects of the level of readership knowledge will affect the documentation, such as:

(a) the purpose of the documentation (tutorial, reference, diagnostic, etc.);

(b) who will read the documentation;

(c) what others sources of information are available to the reader (demonstrations, training, visual aids, etc.);

(d) degree of detail and complexity required;

(e) the reader's job;

(f) the age and experience of the reader.

It may also be helpful to classify readers; to pigeonhole them into groups that relate them according to their needs from the documentation. Different parts of a single publication may need to address members of different classes, so this is particularly important when planning the structure and approach of the documentation. Typically, readers will fall into one or more of the following classes:

(a) laypersons (reading material outside their own field of experience and knowledge);

(b) operators (those who need to carry out some task or achieve a certain result with the aid of the documentation);

(c) decision-makers (those who need to be able to make a decision on the basis of the information supplied);

(d) skilled or expert readers (those who need specialist or detailed information to carry out their work, such as service engineers, training personnel, repair technicians, etc.).

The author may wish to establish to what extent the reader will be familiar with previous documentation or what experience they have of, say, a product accompanied by a manual you have already provided. For example, where a user of a software product is being issued with an update to the existing product, it may be reasonable to assume that their previous experience of the software and its user manual affords them a higher level of subject knowledge. Therefore, the documentation accompanying the software update can be constructed on this assumption.

Similarly, when supplying a document that describes a component of a system, it may be fair to assume that reader will have knowledge of some other part of the system. Again, using an example from the field of computer software, most manuals that describe the installation and use of application software on a computer assume that the reader is familiar with the operation of their computer at some level. It is not necessary to describe what a disk drive is, where it is located on the computer and other aspects of the system that may have been covered by documentation that accompanied the personal computer when it was acquired.

Above all, it is important to understand why a reader will turn to the documentation. The answer to that question should enable the author to determine what kind of information should be included. It should also help establish the order of sections in the document since, logically, they should address the needs of the reader.

As many technical authors already know, few people read technical documentation until they have to; usually when there's a problem or the operation of some equipment or software is not obvious. Therefore, it is important to ensure that a reader can find out what is wanted and understand the information quickly. If the instructions that accompany a video recorder simply explain each of the functions and switches in turn, without regard to what the user may wish to achieve, the usefulness of the document will be limited. The reader will probably turn to the operating manual for answers to such questions as 'how do I set the recorder to tape a programme on channel X that will be transmitted when I am out?'. Analysing the needs of the user in this way determines the content. Continuing this example, one might find that the operator's manual for a video recorder can be organised under the following headings:

Connecting the Video to the TV
Tuning the TV to the Video Channel
Setting the Date and Time
Playing Prerecorded Tapes
Recording a Programme While Watching It
Recording a Programme While Watching Another
Programming the Video to Record While You are Out

and so on.

The headings will mean something to the reader because they relate to what information is needed. Documentation structured in this way is know as 'task-oriented' documentation and where readers fall into the operator or laypersons class, this is usually the best approach. A completely different structure would be required for the service engineer's video repair manual, though it may still be possible to include sections that are task-oriented.

The correct assessment of the readership is probably one of the most important contributions to successful technical documentation. The point was highlighted in a letter to a daily newspaper, in which one individual, clearly irritated by the failure of technical documentation to give due regard to the reader, wrote: 'The only people who manage to get machines to work every time are children, who read the instructions without preconceived ideas. Manuals should be written by teenagers. They are accustomed to working for teachers who need to understand clearly what it is they are writing about'.

COMMUNICATION SKILLS

The technical author must possess good communication skills. These fall into two categories: the skill to communicate clearly through the author's writing and the ability to communicate with others while carrying out his or her duties. Good communication channels are one of the most important tools of a technical author. The following briefly describes the two categories.

A Command of English and Grammar

It is fair to assume that the technical author should be a good communicator of English. A command of good English grammar is certainly an advantage, but the possession of such a skill does not in itself guarantee good technical documentation. Few complaints are filed from users of documentation about the quality of English if it is poor; they are more concerned with the accuracy of the technical content and whether it informs the user as required. However, good English is one of the most important qualities to ensure that the author produces useful and respected documentation. A well-written and clearly presented piece of writing which is easy for the reader to comprehend can only be written with grammatical accuracy. Sloppy English can positively hinder the reader's concentration and any sentence or paragraph which has to be read more than once because of poor sentence structure is no more helpful than inaccurate or misleading data. Misspellings in particular, can be distracting and suggest a lack of attention to detail. They can imply that a similar lack of attention to detail has been given to the technical content of the document.

Even the most well-organised technical document, wholly accurate in the detailed information it conveys, will fail to satisfy the reader who is distracted by other aspects if they are poorly presented. The selection of typestyles, the page layout, the choice of headings, etc., will all affect the ability of the text to communicate its message.

Many articles are written about writing styles, eager to inform authors about the best use of certain words, instructing on the correct use of noun clusters, prepositional phrases, passive voice, jargon and so on. These are often informative and help focus attention on maintaining good communication skills. They are particularly relevant where documentation may have to be translated into other languages and where health and safety aspects are of paramount importance. In these cases, ambiguities that arise from the way a sentence is structured can result in misinterpretation with serious consequences.

Much can be learnt about improving writing techniques from books and articles and by the use of grammar and English-usage references. However, the author should never loose sight of the prime objective of the documentation, which is to communicate information to the reader easily. If too much time and effort is spent analysing sentence structure and syntax where it is not appropriate or necessary, it is easy for an inexperienced author to produce even more incomprehensible documentation. The document should never fail to make simple sense to the reader. As Voltaire pointed out 'Every kind of writing is good save that which bores'.

With the advent of computer-based grammar and style checkers (refer also to Chapter 5), the author can check the text of a word-processed document for grammatical errors and improve writing style as a result. Care must be taken not to take suggestions in either books or grammar checkers too literally. It is quite possible to accommodate all recommendations about writing style and sentence structures suggested by such sources and end up with a far less readable and helpful document. The author should not edit a sentence on the basis of a recommendation that he or she does not understand.

The modern technical author has to consider the best means of communicating the message. Written technical documentation may not be the best form of communication for the job, and no amount of grammar skills will help compile an effective video image if something needs to be conveyed graphically. The author's own skill may therefore need to be extended to suit the medium used for communication. This may mean acquiring skills in graphics and audiovisual technologies.

Communicating with Others

The author will need to obtain information from, and exchange data with, all those who contribute to, or benefit from, a documentation project. If technical staff are involved in providing essential technical data or proofing drafts, the author must be able to communicate with them at their level. As mentioned earlier in this chapter, the author is often the 'bridge' between those of technical expertise and the layperson.

It may be important for the author to be able to communicate with all levels of personnel in his or her own organisation. Setting up project-based meetings on a regular basis can be a good way of monitoring the progress of a job. This may mean that the author needs to communicate with other managers, technicians, trainers, support and maintenance staff, etc.

Where external resources are used in the production cycle of a documentation project, the author may be required to communicate with suppliers and subcontractors. This may include meetings with printers, graphic artists, subcontracting authors and the like. Where documentation is being prepared as an element of a sales contract, the author may well have to be involved with discussions and meetings with clients and customers.

Where an author is expected to attend meetings, he or she should be suitably briefed and ensure that, if called upon to furnish other attendees with information, such as progress reports, all the appropriate information is to hand. If the author is expected to initiate meetings with others, such as engineers, product developers, etc., an agenda should be prepared and forwarded to those concerned in advance. The meeting should have clear objectives. Minutes should be taken and the author must be prepared to chair the meeting.

The communication skills of the author when dealing with others are no more special that those needed by other individuals at a similar level in an organisation. However, given the professional expectations of technical authors, they should set standards by which others should follow. They are, after all, professional communicators.

ORGANISATIONAL SKILLS

Authors whose responsibilities extend beyond that of simply writing technical documentation will need to exercise organisational skills. If the author is responsible for employing and managing resources in the course of duty, it will be necessary to have experience in basic management and commercial activities. The author who understands some of the qualities and skills of good management will be at an advantage. This is particularly true in smaller organisations where authors are often not well managed. This is because managers do not understand the role of the author sufficiently well since few have experienced the role of the technical author. Most managers judge it entirely on their own local perception.

Authors need to be aware of their responsibilities to their employers. Many authors who work in organisations without the supervision of a publications manager can be left out on a limb. They may find it difficult to get adequate recognition and reward for their status and abilities. They may also find it hard to get financial or other resources to help the workload or improve performance and quality. In order to combat this, the author must be prepared to take the initiative. This means being prepared to provide information to superiors about workloads, resource requirements and cost analyses.

Authors working alone or without direct supervision may need to be able to:

(a) schedule all documentation jobs;

(b) suggest ways in which performance can be improved;

(c) devise means to monitor job progress and evaluate performance;

(d) take effective action to improve quality, performance and economy;

(e) procure necessary resources for the job;

(f) maintain an effective job control system, including cataloguing, filing and analysis of documentation jobs.

Since these skills are more relevant to the publications manager, they are covered in more detail in Chapter 9, Publications Management. However, all professional technical authors should be prepared to undertake responsibilities such as planning, budgeting, resource management, etc. and these should form part of the author's training.

SUMMARY

1. The author's role is as diverse as the different fields of technology in which he or she works. The activities can extend from simply the preparation and writing of texts to author/manager responsibilities that encompass those of a publications manager.

2. Software technical authors are usually producing documentation which is as much a marketing aid as it is a useful user guide. They often have to prepare documentation without any existing specifications and maintain constant dialogue with the software developers. This is particularly important as the documentation cycle often has to end before the development cycle.

3. Hardware technical authors may have greater responsibilities to carry with respect to health and safety concerns than software authors. A technical education and training is almost always essential and many hardware authors carry out documentation as a supplementary duty to other technical activities such as development and maintenance.

4. The role of the author encompasses a number of key areas of responsibility. The main task being to communicate technical information accurately using the appropriate medium effectively. The result should be easy to understand by the readership and suit their needs precisely.

5. The level of subject knowledge required by the author will depend upon the complexity of the content and the purpose for which the documentation is intended. It is unlikely that even a trained technical author possesses all the relevant subject knowledge and must be able to obtain relevant information from appropriate sources.

6. An understanding of the user is important to ensure that documentation is targeted accurately and presents information at the right level of readership. An author may need to carry out an evaluation of the intended readership if it is not obvious.

7. An author must have good communication skills, both in being able to convey clear meaning through his or her writing and by being able to liaise with others involved in the documentation project. Documentation that contains grammatical and spelling errors will result in a loss of confidence by the reader who will question the accuracy of information conveyed.

8. Some organisation skills are important, particularly where the author exercises control of the publication beyond the writing stage and where other resources are employed to assist in the project.

FURTHER READING

Kirkman, J. *Full Marks* (ISTC, 1991).

Turk, C., Kirkman, J. *Effective Writing* (E&FN Spon, 1982).

2 Technical Writing

This chapter discusses the main aspects that affect the process of technical writing from conception of the documentation through to prepress preparation. By their very nature, technical documents, user manuals and the like are judged in their effectiveness by direct relationship to their achievement of the purpose. Since the function of technical documentation is to provide a 'backup' service, which is difficult to evaluate, and because there is a tangible end-product which lends itself to criticism, it is easy for the physical appearance to give a false sense of quality. The quality aspect of documentation is often affected by the degree of planning undertaken. It is also dependent upon the related development process of the product about which the documentation is written. The implementation of the documentation cycle, the correct use of resources, the choice of format and style, all affect the overall quality. No matter what degree of excellence is afforded to the design and production of the finished documentation, lacking in other areas of technical writing will result in its failure to 'deliver the goods'.

PRELIMINARY PLANNING

A good technical document must be well planned. The best place to start is by defining the objective of the documentation. If targeted correctly and provided adequate resource is made available, the objective of the documentation should be achieved without the need to adjust the original plan. In reality, however, concessions and compromises are often made throughout the documentation cycle, and the original intention of the documentation may be affected by many external influences. For example, the time-planned documentation cycle for a particular publication may well be undermined if the product development cycle on which it relies is not coordinated properly. If the development timescale goes adrift, it may reduce the time available to complete certain sections of the documentation, which, as a result, may end up being less accurate and informative.

When planning the documentation cycle, many factors need to be assessed. These include evaluating the available resources and determining how much information will be available to complete various stages of the documentation. In the case of a software product manual, where development specifications and blueprints may be scarce or non-existent, the author will need to decide whether the documentation can begin before completion of the software product itself. There may be opportunity for background work to be done on a user guide, or it may even be possible to write most of the manual on the basis of a suitably detailed system specification if one is available. However, if illustrations of output from the system need to be included in the final publication, it may not be possible to obtain such material until the product is finished. Such factors must be taken into account when planning timescales and deciding upon the content of the finished document.

The documentation project should be planned according to a number of objectives and parameters. These may include:

(a) evaluating the objective of the document;

(b) ensuring sufficient resource is available;

(c) identifying any constraints imposed on the documentation cycle by other factors, such as development, author training on the product, proofing cycles, etc.;

(d) scheduling required resources;

(e) providing mechanisms for modification and updates;

(f) ensuring all international, legal, industry standard and other conventions are adopted;

(g) providing a suitable means of production and distribution.

Table 2.1 shows a checklist for determining the project objectives, requirements and constraints for a technical publication. Means should be provided to ensure that sufficient analysis can be carried out before the documentation work begins. The author must ensure adequate information is gathered to help structure the content of the manual. This includes identifying the users of the documentation, the environment in which it will be used, what kind of information must be included, etc. Table 2.2 shows a checklist for the analysis of a technical publication.

The planning stages need to take into account standards of writing, editing, illustration, and packaging. The timescale and costs will effect the end-product for any technical publication. No manual that can be written with infinite resource. Commercial, development and economical factors all play a part in deciding the limits of production of a technical document. Clearly, all planning must be done in the knowledge of these limits. The author must have some idea of the overall expense that can be afforded. This often conflicts with the estimate of costs to achieve the desired objective. Compromise needs to be made and should form part of the planning process and not be 'evolved' during the documentation cycle itself. Table 2.3 shows an example project objectives planning checklist, while Table 2.4 illustrates a typical documentation plan. All the sample checklists provide a means of recording decisions and action as at the preliminary and planning stages of documentation development. The checklists are reproduced by courtesy of the British Standards Institute. For actual use, a checklist may have to be laid out differently to allow more space to be provided for the recording of relevant information.

Table 2.1 Project objectives, requirements and constraints

		Obtained	*Interpreted*	*Recorded*
Objectives	Product			
	Sales			
	Schedule			
	Usability			
Requirements	Modification			
	Translation and national culture			
	Packaging			
	Legal			
	Standards and conventions			
Constraints	Cost constraints			
	Quality management*			
	Provision of technical information			
	Approval authorities			
	Change of control			
	Availability of resources			

Output	A list of the project objectives, requirements and constraints, including sources for each.	*Reference*

*Quality management plays a key role in documentation planning. The proofing and checking process can have a significant impact on timescales. The author must be aware of the degree of quality control that needs to be exercised over the project. For those involved in publications of a highly technical nature, or where health and safety concerned are paramount, implementation of a quality management system such as that of BS5750, should be considered.

20   TECHNICAL DOCUMENTATION

Table 2.2 Analysis checklist

		Document reference
Inputs	Users	
	Tasks	
	Learning stages	
	Environments	
Activities	Decide what information is needed	
	Decide what kind of information is needed	
	Decide where documentation is to be read	
	Record user's information needs	
	Group information into documents	
	Decide what types of documents	
Output	Outline documentation plan	

Table 2.3 Project activities planning checklist

	Document reference
Editorial standards	
Illustration standards	
Packaging style	
Packaging method	
Costs	
Overall schedule	
	Date and reference
Write documentation plan	
Review documentation plan	
Issue revised documentation plan	

Table 2.4 Documentation plan checklist

Title
Content
Structure
Style
Page numbering
Section numbering
Page layout and typography
Binding
Development method
Production method
Method of updating
Costs
Schedule

FORMATTING

Apart from acquiring the information needed to plan documents, the format and organisation of the contents must be considered. This should be approached in a logical fashion. Structuring the outline of a document can assist the author writing individual components of the work, as they can be related to the whole. Once a clear structure is formulated, individual sections can be worked on by more than one individual, without

22   TECHNICAL DOCUMENTATION

losing the overall format. The structure plan needs to include enough detail to determine what kind of information will be provided in each section of a document to avoid unnecessary repetition, while ensuring there is sufficient information for a user at any stage of reading.

The format of a user manual, for example, must take into account the prerequisite knowledge of the reader and what information is provided in other ways, such as associated training, reference cards, on-screen help, etc. The format of any given piece of documentation is best broken down into the smallest component parts practicable. An example of some of the subjects that should be covered by the contents is shown below for a technical product user manual:

(a) Introduction — provides an overview of the document and explains any document conventions used and level of prerequisite knowledge expected of the reader.

(b) Description of the product — explains what the product is in the form of an overview. Includes any relevant information about the version, manufacture, etc.

(c) Specification — provides a technical specification of the product including the description of the environment in which it should be used, relevant standards adopted and other technical characteristics.

(d) Requirements for use — explains what other components or products are needed to operate the product including the minimum requirements for adequate use.

(e) Getting started — covers the information necessary to begin using the product.

(f) Components supplied — a list of all components of the package so the user can check that all items are present.

(g) Installation — describes the means by which the product is installed for use. This may include assembly instructions, wiring requirements, software installation on a computer, etc.

(h) First-time operation — covers the switching on, starting up, loading or other means of initial operation.

(i) Day-to-day operation — explains how to use the product to achieve the particular objectives expected of the product.

(j) Maintenance — guide to ordinary maintenance of product as may be expected to be done by the user.

(k) Troubleshooting — help on possible problems that may arise in the normal use of the product.

Different objectives and requirements will be identified for other kinds of documentation. The format of a reference manual will be different to that required of a tutorial or a task-oriented manual. In all cases, the document structure should ensure the contents provide all necessary information to the user to achieves the objectives.

LANGUAGE AND STYLE

All technical documentation tries to 'teach' or inform the reader. The differences are between those that teach well and those that teach badly, and between those that teach valuable things and those that teach irrelevant or trivial things. An author who does not care about how he or she writes is just as likely to be a bad author as one who does not care about the subject or information being conveyed. Good communication through documentation, like good design, begins in the mind. Certain attitudes and thought processes will precede the words and a good communicator will always consider the recipient of the message: so must an author consider the reader when forming the text through which the message is being conveyed. The form and style adopted in the writing must be suited to the purpose of the documentation. Form and style should not simply adhere to rules or regulations laid down by some preconceived notion about technical documentation. Except for cases where documentation must conform to some British Standard or specification, the author's job includes the responsibility of considering and implementing the most suitable form and style for the job. This may be easier for authors working inside organisations for whom the documentation is being prepared. Contract technical authors may be forced to adopt unsuitable forms and styles because they are demanded by the client; perhaps because they conform to an existing house style. In these circumstances, the author can only politely and diplomatically advise about, but not necessarily control, the application of a suitable form and style.

There are tried and tested approaches for preparing certain types of documentation and qualified technical authors will have encountered these as part of their own career training. Individual style can be build on sound principles, but the author should be wary of trying to be innovative in every publication. Where a document forms part of a series, in may be more important to maintain a consistent approach to form and style than introduce change at an inappropriate point.

Writing began as a specialist skill. Today, good writing remains so. Since becoming a widespread form of communication among all walks of life, in an overwhelming variety of forms and styles, from academic literature to 'gutter press', the English language has come under tremendous stress and undergone, and survived, considerable abuse. In industry and commerce, where new technology and progress are stretching the vocabulary of the English language, one sees newly invented terms and expressions, jargon and 'empty' words being invented at an astonishing rate.

Those who work in technical environments encounter many technical words and terms, many of which are unintelligible to the layperson. Assessing the level of readership was discussed in Chapter 1, and it is essential that the author determines the appropriate level of vocabulary to be used when communicating through documentation. So often, technical jargon is used where more common words are adequate and usually preferable. There are countless examples and those authors who work in a high-technology environment may find it difficult to remain constantly aware of the specialised nature of the terms they use every day. Some particularly bad examples can be found in software documentation, much of which arises for the poor standards of English that is presented to the user on the screen while an application is in use. Here one finds such ugly words and phrases such as 'pressing F5 will abort the function'; 'pressing Esc returns control to the previous menu'; 'click on the delete button to kill a file'; 'the program uses dynamically expanding file structures'. The constant use of such verbiage can often result in quite bizarre and confusing sentences and phrases appearing in documents such as 'press Enter to exit' and 'drag the folder with mouse into the wastepaper basket'.

The author should be ruthless in the use of accurate and precise terms and descriptions, but, above, should keep the text simple. This is particularly difficult for authors writing for technical audiences. Just because the readership happens to possess a high level of subject knowledge is no excuse for being unnecessarily wordy or using complex language. The following reminder about this subject was pinned to a computer belonging to a colleague of mine:

> 'Such preparations shall be made as will completely obscure all Federal buildings occupied by the Federal government during an air raid for any period of time from visibility by reason of internal or external illumination. Such obscuration may be obtained either by blackout construction or by terminating illumination. This will, of course, require that in building areas in which production must continue during a blackout, internal illumination be provided, that construction may be provided. Other areas, whether or not occupied by personnel, may be obscured by terminating the illumination'.

Rewrite by Franklin D. Roosevelt:

> 'In buildings where work will have to keep going, put something across the windows. In buildings where work can be stopped for a while, turn out the lights'.

Technical documentation must be complete and specific about what it is to convey. When documentation a fault-finding procedure, one should consider the questions that may be asked by reader when trying to establish what action to take rather than simply listing answers to problems. It helps to expand text beyond simply a technical specification by considering what the reader is trying to achieve. For example, if a document is intended to describe the installation procedure for a piece of equipment, apart from covering all the tasks necessary for installation itself, it may be appropriate to cover what action the user must take in the event of unexpected results. See the examples in Figures 2.1 and 2.2.

Apart from being complete, documentation should only contain information which is relevant. There is no point in including detailed fault diagnostics in a user manual if, as a result of certain failures, the system can only be serviced by a qualified engineer; such information would be better reserved for a separate document for the engineer. Irrelevant material included in a document disrupts the flow of reading.

Much fuss is made about the most suitable style of writing for technical documentation. Probably the most commonly expressed preference is the use of the active voice rather than the passive. For example, the following sentence is written in the passive voice:

> 'You can print a report showing the current settings in the system'.

TECHNICAL WRITING 25

Remove the system components from the packaging and place on a flat, even surface. Connect one of the power cables supplied to the power socket at the read of the processing device and the other to the power output of the VDU. The cable connected to the VDU should then be connected to the port marked 'video' on the processor. The keyboard connects to the socket marked 'KB' on the rear panel.

Switch the system on. The device will start a warm-up procedure during which self diagnogstics will check that the system is functioning correctly. Any errors will be reported on the VDU during this process. Assuming that all is working well, the screen will display the message 'insert system disk in drive. Press any key when ready'. Follow these instructions and wait for the system menu to be displayed.

Figure 2.1 Example text.

1. You will need a flat, even surface to assemble the components of your system and access to two power sockets nearby. No tools are required for assembly.

2. Remove the system components from the packaging. You should find the following items:
   - Processing device
   - VDU (with cable attached)
   - Two power cables
   - Diskette labelled 'System Disk'
   - Keyboard (with cable attached)

   If anything is missing, report the matter to your supplier immediatly.

3. Connect the components' cables as follows (all connections are made on the rear panel of the processor):
   - one power cable into the power socket of the processor;
   - the other power cable to the socket on the VDU;
   - the free end of the cable attached to the VDU into the video output port marked 'video' on the processor;
   - the keyboard's cable into the socket marked 'KB'.

4. Make sure the power cables are then connected to a mains supply before switching on the device. Once switched on, the device should begin its warm-up procedure. You will hear the cooling fan at first, then a beep, as the start-up process begins. If nothing happens, there may be a fault with the equipment, the cables, fuses may have blown either on the power cable plugs or in the rear panel of the device, or you may have simply omitted to switch on the mains supply.

Figure 2.2 Example text, improved format.

The following sentence, on the other hand, is written in the active voice:

'Print a report to show the current settings in the system'.

The latter is more direct. As for whether it is an improvement is subjective. There is no sound evidence that writing in the active voice improves the effectiveness of technical documentation. Where documentation is not likely to be translated into other languages, the most important thing is to ensure the reader understands what has been written.

EDITING

Editing texts and other elements of technical documentation forms a significant part of the workload. Editing can be carried out by any competent individual provided it is within the scope of a brief and that the individual concerned has sufficient authority, in terms of subject knowledge. The editor must be aware of the terms of reference in which the text is to be edited. If asked to check the technical accuracy of facts because he or she has the information or knowledge available to do so, that editor's task should be restricted to that function only. While it is appropriate for the editor to point out any weaknesses that are not in his or her terms of reference, care must be taken to avoid subjective amendments.

The author or publications manager may wish to compile a check-list of items for the editor. When the editor checks through the document, he or she is alerted to the points that are most relevant. The checking of spelling and grammatical aspects of a technical document should not be left in the hands of an editor whose job it is to check technical content, unless that individual also happens to be suitably qualified in English.

The responsibilities of the editor must be relevant to the individual's skills and abilities. At all times, the editor must avoid subjective criticism if he or she is to gain respect of the author and to be able to exercise authority. Authors are prone to being oversensitive to criticism of their work and the editor can often have a difficult task simply persuading an author to accept any changes to the text; especially where there is an element of subjective opinion involved. To avoid conflicts, editors must be clear of their responsibilities, and to know the limits of their editorial powers. Furthermore, it important that all other individuals involved in the publication are aware of the editor's powers.

Editors who are not trained authors should be wary of adjusting the organisation and structure of document prepared by a qualified technical author. A design engineer would not take kindly to a technical author adjusting the design of a piece of equipment. It is commonplace, however, for all individuals to believe that, since we all read and write, anyone is qualified to criticise a document, even though it may have been professionally designed and written.

Before any editing can be done, the editor must have a clear framework in which to work. The terms of reference for the editor should be discussed with the author and publications manager. If documentation is being produced by a contracting service, this aspect is even more critical; for example, if the editor suggests subjective changes to a document that is being originated by a subcontracted author, additional and unnecessary costs and delays in the project may be incurred. The terms of reference must be based on sound objectives and the following points should be considered in this respect:

— is there a specification to which the documentation must conform?

— must the document follow a house style?

— is there an agreed standard for terminology?

— what is the prime objective of the document?

— what kind of document is it; for example, a tutorial, troubleshooting guide, reference manual, assembly instructions, etc.?;

— what is the readership of the document and what level of knowledge is assumed?

PROOF-READING

There are two kinds of proof-reading. Traditionally, it involves the reading of both an original and a copy and ensuring there are no differences between the two. This is important where the production process for a document involves the replication of text and/or illustrations, for example, when a typescript is rekeyed into a phototypesetter. The second kind of proof-reading is really proof-*checking*, though the former term is the most widely used. This involves ensuring that the content of the document is accurate, that it conforms to appropriate specifications and that it reads well. Modern methods of documentation production have tended to render the traditional kind of proof-reading redundant. Copy is invariably prepared on word processors, and rekeying is eliminated by transferring text electronically from one part of the publication system to another.

All documentation should be proof read and in most circumstances, there will be a number of stages during the production cycle at which proof-reading needs to be done. The nature of the proof-reading itself will probably vary according to which stage the documentation is at. During the initial stages, when text and illustrations are drafted, a general read-through may be all that is necessary to determine whether the material covered is appropriate and relevant. As the documentation project progresses, greater attention will need to be paid to detail, such as spelling and grammar and whether it conforms to any house styles and specifications. At the final prepress stage, very careful attention will need to be paid to the copy, cross-references, pagination and indexing.

The persons responsible for proof-reading will depend upon the type of documentation being produced. Wherever possible, more than one person should be involved in the proof-reading; it is most important that the person who originated the material is not the only proof-reader. Authors and illustrators are generally not good at proofing their own material. However, those who are employed to do it should be suitably qualified. A technical document may require vetting by a person familiar with the technical nature of the content, but it may be necessary for another individual to check style and grammar. When marking proofs with corrections, it is important that the are made clearly and without ambiguity. It is a good idea to use a different colour on hard-copy proofs, such as red, to make the mark-ups stand out clearly. The person responsible for subsequently correcting the proofs should be in no doubt about what change is to be made.

If proof-reading or checking is being carried out by more than one individual, there should be an agreed method of marking changes so that the same mistake is marked up in the same way each time. Where proof-reading is carried out by in-experienced and unqualified individuals, there is a tendency for them to adopt a condescending approach by writing critical comments against text which are largely subjective and unhelpful to the author. There is no point writing words like 'Confusing' or 'What?' against a paragraph and expect an author to understand what it is that checker want amending.

Where no accepted methods of proof correcting are in place, it is perhaps worth considering adopting the British Standard on proof correcting BS5261: Part 2 1976. An extract is reproduced on the following two pages by courtesy of the British Standards Institute.

SPECIFICATIONS

Specifications for technical documentation determine the content and format to which particular manuals or other documentation are to be produced. Though almost non-existent in the computer software industry, in hardware, engineering, aviation, defence and other technology areas, specifications are often used to ensure a uniform standard of work. The Ministry of Defence commission many technical publications all of which are, without exception, written to meet a specification. Where documents are to be written to meet the criteria specified, authors have a responsibility to understand an implement the demands of the specifications involved.

Authors already working in the areas of technology mentioned above may already be familiar with certain specifications. Indeed, many jobs or commissions for technical writing in aerospace or defence environments make knowledge of the relevant specifications a prerequisite. There are a number of British Standard specifications for particular type of publication and industry. Probably the most important of which is BS4884. There are, however, many pan-European and international standards that the author may encounter, such as ATA 100, produced by the Air Transport Association of America, produced as a specification for manuals relating to maintenance and operation of aircraft. In the software industry, BS7649, which covers guidelines on the preparation and production of user manuals for application software, is the only significant British Standard at present.

Writing style, the format and presentation of content and other aspects of technical documentation production that have to meet particular standards are not covered in this book, since the standards themselves dictate what is required. A list of some of the relevant specifications and standards is given Chapter 12.

TECHNICAL WRITING 29

Table 2.1 Proof correction marks

Note: The letters M and P in the notes column indicate marks for marking-up copy and correcting proofs respectively.

Group A General

No.	Instruction	Textual Mark	Marginal Mark	Notes
A1	Correction is concluded	None	/	P Make after each correction
A2	Leave unchanged	------ under characters to remain	✓ (encircled)	M P
A3	Remove extraneous marks	Encircle marks to be removed	×	P e.g. film or paper edges visible between lines on bromide or diazo proofs
A3.1	Push down risen spacing material	Encircle blemish	⊥	P
A4	Refer to appropriate authority anything of doubtful accuracy	Encircle word(s) affected	(?)	P

Group B Deletion, insertion and substitution

B1	Insert in text the matter indicated in the margin	∧	New matter followed by ∧	M P Identical to B2
B2	Insert additional matter identified by a letter in a diamond	∧ ∧	∧ Followed by for example ⟨A⟩	M P The relevant section of the copy should be supplied with the corresponding encircled letter marked on it e.g. ⟨A⟩
B3	Delete	/ through character(s) or ⊢—⊣ through words to be deleted	∂	M P
B4	Delete and close up	⌒/ through character or ⌣ through characters e.g. chara̸cter cha̲ra̲a̲cter	⌒∂	M P

30 TECHNICAL DOCUMENTATION

No.	Instruction	Textual Mark	Marginal Mark	Notes
B5	Substitute character or substitute part of one or more word(s)	/ through character or ⊢——⊣ through word(s)	New character or new word(s)	M P
B6	Wrong fount. Replace by character(s) of correct fount.	Encircle character(s) to be changed	⊗	P
B6.1	Change damaged character(s)	Encircle character(s) to be changed	✗	P This mark is identical to A3
B7	Set in or change to italic	———— under character(s) to be set or changed	⊔⊔	M P Where space does not permit textual marks encircle the affected area instead
B8	Set in or change to capital letters	≡≡≡ under character(s) to be changed	≡	
B9	Set in or change to small capital letters	═══ under character(s) to be set or changed	=	
B9.1	Set in or change to capital letters for initial letters and small capital letters for the rest of the words.	≡ under initial letters and ═══ under rest of the word(s)	≣	
B10	Set in or change to bold type.	〰〰 under character(s) to be set or changed	〰	
B11	Set in or change to bold italic type.	——— 〰〰 under character(s) to be set or changed	⊔⊔ 〰	
B12	Change capital letters to lower case letters	Encircle character(s) to be changed	≢	P For use when B5 is inappropriate

TECHNICAL WRITING 31

No.	Instruction	Textual Mark	Marginal Mark	Notes
B12.1	Change small capital letters to lower case letters	Encircle character(s) to be changed	≠	P For use when B5 is inappropriate
B13	Change italic to upright type	Encircle character(s) to be changed	4	P
B14	Invert type	Encircle character to be changed	↺	P
B15	Substitute or insert character in 'superior' position	/ through character or ∧ where required	Y under character e.g. ³	P
B16	Substitute or insert character in 'inferior' position	/ through character or ∧ where required	∧ over character e.g. ½	P
B17	Substitute ligature e.g. ffi for separate letters	⊢──⊣ through characters affected	⌒ e.g. ffi	
B17.1	Substitute separate letters for ligature	⊢──⊣	Write out separate letters	P
B18	Substitute or insert full stop or decimal point	/ through character or ∧ where required	⊙	M P
B18.1	Substitute or insert colon	/ through character or ∧ where required	⊙	M P
B18.2	Substitute or insert colon	/ through character or ∧ where required	;	M P

32 TECHNICAL DOCUMENTATION

No.	Instruction	Textual Mark	Marginal Mark	Notes
B18.3	Substitute or insert comma	/ through character or ⋏ where required	﹐	M P
B18.4	Substitute or insert apostrophe	/ through character or ⋏ where required	ʾϒ	M P
B18.5	Substitute or insert single quotation marks	/ through character or ⋏ where required	ʿϒ and/or ʾϒ	M P
B18.6	Substitute or insert double quotation marks	/ through character or / where required	❝ϒ and/or ❞ϒ	M P
B19	Substitute or insert elipsis	/ through character or ⋏ where required	•••	M P
B20	Substitute or insert leader dots	/ through character or ⋏ where required	⊙	M P Give the measure of the leader when necessary
B21	Substitute or insert hyphen	/ through character or ⋏ where required	⊢⊣	M P
B22	Substitute or insert rule	/ through character or ⋏ where required	⊢⊣	M P Give the size of the rule in the marginal mark e.g. ⊢1 em⊣ ⊢4 mm⊣

TECHNICAL WRITING 33

No.	Instruction	Textual Mark	Marginal Mark	Notes
B23	Substitute or insert oblique	/ through character or ⋋ where required	⊘	M P

Group C Positioning and spacing

No.	Instruction	Textual Mark	Marginal Mark	Notes
C1	Start new paragraph	⌐┘	⌐┘	M P
C2	Run on (no new paragraph)	⌒	⌒	M P
C3	Transpose characters or words	⎵⎴ between characters or words, numbered when necessary	⎵⎴	M P
C4	Transpose a number of characters or words	3 2 1 |||	1 2 3	M P. To be used when the sequence cannot be clearly indicated by the use of C3. The vertical strokes are made through the characters or words to be transposed and numbered in the correct sequence.
C5	Transpose lines	⌐	⌐	M P
C6	Transpose a number of lines	─── 3 ─── 2 ─── 1		P. To be used when the sequence cannot be clearly indicated by C5. Rules extend from the margin into the text with each line to be transposed numbered in the correct sequence.
C7	Centre	⌈enclosing matter to be centred⌉	[ ]	M P
C8	Indent	⌐┘	⌐┘	P. Give the amount of the indent in the marginal mark

34 TECHNICAL DOCUMENTATION

No.	Instruction	Textual Mark	Marginal Mark	Notes
C9	Cancel indent			P
C10	Set line justified to specified measure	and/or		P Give the exact dimensions when necessary
C11	Set colunm justified to specified measure			M P Give the exact dimensions when necessary
C12	Move matter specified distance to the right	enclosing matter to be moved to the right		P Give the exact dimensions when necessary
C13	Move matter specified distance to the left	enclosing matter to be moved to the left		P Give the exact dimensions when necessary
C14	Take over character(s), word(s) or line to next line, column or page			P The textual mark surrounds the matter to be taken over and extends into the margin
C15	Take back character(s), word(s), or line to previous line, column or page			P The textual mark surrounds the matter to be taken back and extends into the margin
C16	Raise matter	over matter to be raised / under matter to be raised		P Give the exact dimensions when necessary. (Use C28 for insertion of space between lines or paragraphs in text)
C17	Lower matter	over matter to be lowered / under matter to be lowered		P Give the exact dimensions when necessary. (Use C29 for reduction of space between lines or paragraphs in text)
C18	Move matter to position indicated	Enclose matter to be moved and indicate new position		P Give the exact dimensions when necessary

TECHNICAL WRITING 35

No.	Instruction	Textual Mark	Marginal Mark	Notes
C19	Correct vertical alignment	∥	∥	P
C20	Correct horizontal alignment	Single line above and below misaligned matter e.g. m<u>i</u>sa<u>li</u>gned	=	P The marginal mark is placed level with the head and foot of the relevant line
C21	Close up. Delete space between characters or words	linking ⌢⌣ characters	⌢⌣	M P
C22	Insert space between characters	| between characters affected	Y	M P Give the size of the space to be inserted when necessary
C23	Insert space between words	Y between words affected	Y	M P Give the size of the space to be inserted when necessary
C24	Reduce space between characters	| between characters affected	T	M P Give the amount by which the space is to be reduced when necessary
C25	Reduce space between words	T between words affected	T	M P Give amount by which the space is to be reduced when necessary
C26	Make space appear equal between characters or words	| between characters or words affected	Ȳ	M P
C27	Close up to normal interline spacing	(each side of column linking lines)		M P The textual marks extend into the margin

HOUSE STYLES

Unless you are working on technical documentation which has to conform to British or other standards, you should consider defining a 'house' style. By doing so, different authors working on documentation for a particular project will have a common point of reference that will ensure that their work has a consistent approach and style, which is important when all the parts come together as a whole. It is also important for organisations that employ the services of contract authors who will need guidance on what aspects of design and writing style you employ.

The following are some aspects of documentation design that may be included in a house style specification.

Page Style

Determine the layout styles to be used for those elements of the documentation where it is practical to adhere to a format. You may require different layouts for different components, such as user guides, repair manuals, technical specifications, support literature, etc. The specification for a page style should include measurements for paper, margins, indents, and the position of various elements on the page.

Headings

Specify how headings are to be used in the documentation. Consideration should be given to levels of heading will begin a new page, what relative size and style of typeface should be used, whether the headings are to be numbered and to how many levels, etc.

Spelling and Hyphenation

It is important to identify preferences for spellings where alternatives are possible. For example, 'specialise' may be preferred to 'specialize', 'organise' to 'organize', 'indexes' to 'indices', etc. Many hyphenated words are also compound words, such as 'sub-division' and 'subdivision'. Both are acceptable, though the less hyphens used, the better. Some words look better hyphenated, particularly where two vowels come together as in 'cooperate' which reads better as 'co-operate'. The house style should indicate preferences to ensure consistency.

Punctuation

Some aspects of punctuation may need to be clarified, such as the use of full stops after certain abbreviations like 'etc' and whether single or double quotes are used. It is important to ensure that the use of punctuation when giving instructions on keying in command on a computer may be misinterpreted. Make sure that there is clear distinction about what the user must type. For example, if the text reads "type 'dir' at the system prompt", it is unclear whether the single quotes should be keyed as well as the letters dir. The house style should establish clear rules in this respect.

Capitals

Capital letters are often used abundantly in technical documentation for no apparent reason. In the case of software user documentation, this approach is frequently used in screen layouts and on messages and prompts. As a result, they tend to get duplicated in the manual in the same way. For example, 'Press Space when the Drive Letter is highlighted, then insert the Disk into the chosen Drive' uses capitals in an unnecessary way, whereas 'choose the Backup Data Files option from the menu' at least distinguishes the name of an option on the screen. The house style should set guidelines with some examples for the author to follow.

Text Highlights

Guidance should be given on the use of text highlights such as bold, italic, underline, etc. For example, it may be appropriate to use italics for variables and bold for warnings or important text. Since these attributes can be used to denote so many things, it is best to restrict their use to specific cases.

Abbreviations and Units

A decision should be made as to whether abbreviations are to be used with or without full stops. Abbreviations should consistent with their recognised use, for example MB for megabyte, Mb for megabit. SI units should follow British Standard, for example mm, m, km (but not cm), g (but not gm), etc. A house style should also identify any special abbreviations used in the organisation's own documentation but not necessarily recognised elsewhere.

Dates

The format of dates used in documentation should be specified in the house style. For example, dates may have to be in full, such as 22nd July 1994 or follow conventions used in the system, such as in software application documentation, where dates may be entered by users of the system in an abbreviated format, such as 22-07-94.

Notes and Reference System

How footnotes and references to other publications are handled should be included in the house style, particularly in the case of technical reports, journals, etc.

Tables, Lists, Figures, Maps, Plates, etc.

It is commonplace for tables, lists, figures, maps, plates, etc. to be referenced within the text and some kind of numbering system needs to be specified. For example, it may be appropriate to give tables and figures separate numbering sequences. These may be simply a single sequence, such as 1, 2, 3, 4, etc., throughout the document and regardless of chapter, or within chapters with the chapter number as a prefix, such as 1.1, 1.2, 1.3, then 2.1, 2.2, 2.3, etc.

Bulleted or indented lists may require numbering, especially if the sequence of items in the list are important. It is commonplace to number steps of an operation that must be carried out in a particular order, for example, 1, 2, 3, etc. It may also help to refer to particular steps in a process by referring to the number in the list. If the sequence itself is not important, other item references may be used such as (a), (b), (c), or roman numerals like (i), (ii), (iii), etc.

Copyright and Trademarks

Where quotes from other works are included in a document, the house style should specify how such quotes are acknowledged and how trademarks are similarly acknowledged. In some cases, where referenced to trademarks are few, it may be possible to acknowledge them in the text with the appropriate symbol, for example TimeSaver®, or if there are many occurrences, it may be preferable to acknowledge all trademarks among the leading pages of the document.

Indexing

There is a British Standard document BS3700 which gives information about the preparation of indexes. If any particular methods or structures are employed for a particular family of documents, the house style should provide information and give examples. The presentation and nesting of entries in the index can be a complex subject in itself. For example and entry such as 'RS232 serial interface', may appear in the index under 'R' as is, under 'S' with an entry of 'Serial interface, RS232', under 'P' for 'Ports' and or 'Printers', such as 'Port, serial, RS232' or maybe under all such possibilities for a comprehensive approach.

PREPRESS PREPARATION

Prepress preparation refers to any aspect of technical documentation production that affects the way in which the document will be published. For example, if a document is going to be photocopied, this may mean production of master pages for the purpose. If a more complex production method is to be employed, such as litho printing, the camera-ready artwork may be prepared in a number of ways. Where phototypesetting is involved, the author may be required to provide disks of the text files in a suitable format. If the text is to be interfaced with a page make-up system, there may be some generic coding of the text required or the adoption of certain formatting conventions to ensure that appropriate levels of heading, text and other characteristics are reproduced accurately.

The way in which a technical document is structured and the planning of its development is therefore affected by prepress preparations. The following is a checklist of some of the aspects of technical documentation prepress preparation that may affect the process of technical writing:

(a) the method of artwork production (for example, word-processed, photocopied, desktop publishing, phototypesetting, etc.);

(b) the system used for artwork production;

(c) the author's standard method of coding for prepress interfacing;

(d) the method of illustration;

(e) the integration of graphics and other objects in the document;

(f) the method printing and finishing.

The choice of system used to produce the technical documentation is important as far as prepress preparation is concerned. The author should be aware of the various technologies available for printing and document production. For example, with the advent of digital presses and systems such as Xerox's Docu-Tech, it may be relevant for the technical documentation to be prepared on a system that can be interfaced to such technologies directly. If a document is to be phototypeset, production costs will be cheaper if the document can be prepared on a system that can be either directly connected to a phototypesetting system or transferred electronically from the source application to the typesetting system.

If more traditional methods of documentation production are to be employed, such as cut and paste in a design studio, the author may have more flexibility over the choice of system which will be used to original the document and the technical writing process may involve less prepress preparation.

SUMMARY

1. Planning of a technical documentation project is a fundamental part of the writing process. This involves evaluating the objective of the material, ensuring the structure and content fulfil user needs and meets required standards. Resources must be scheduled in accordance with proposed timescales which must be integrated with the development process where applicable.

2. The format of the documentation must be appropriate for the purpose. Different types of documentation require a different structure.

3. Form and style of writing varies greatly. There are established guidelines for good writing styles, but some approaches are a matter of preference. The most important factor is to ensure the content of the document is easily understandable by the user.

4. Editing needs clear terms of reference in which to work. The responsibilities assigned to editors of technical documentation should be commensurate with their skills and abilities.

5. Although proof-reading is traditionally the comparison between an original and a copy document, the term is now generally used to refer to proof-checking.

6. When proofs are marked-up for correction, a consistent approach to the use of correction marks should be adopted, particularly where proofs will be checked by more than one individual.

7. The format, form and style of writing in some cases is governed strictly by recognised standards and specifications. For example, technical documentation commissioned for the Ministry of Defence and aviation organisations have to follow specific standards.

8. Where no particular specifications apply, it may be useful to draw up a house style — a simple specification that ensures documentation produced for the same organisation follows a consistent approach to certain details, such as the use of terminology, abbreviations, heading levels, etc.

9. The way in which technical documentation projects are planned can be affected by the method employed for publication. The work that must be done to make the documentation ready for reproduction is known as prepress preparation. The author must choose a suitable means of preparing documentation that benefits the chosen method of production.

FURTHER READING

Austin, M. *Technical Writing and Publication Techniques* (ISTC, 1987).

Kirkman, J. *Full Marks* (ISTC, 1991).

Turk, C., Kirkman, J. *Effective Writing* (E&FN Spon, 1982).

3 Technical Publications

Different types of technical publications require different approaches to design and production. They are also intended to achieve different objectives. In each case, the author must decide on what kind of information needs to be communicated, then choose the most appropriate type of publication to achieve this. This chapter discusses some of the main types of technical publication, though there can be many variants on the main types and a great deal of technical documentation tends to be a hybrid between more than one type.

It should be stressed that, in the following sections, emphasis is on hard-copy publications. However, when choosing a suitable form of communication, the author may find that some other medium is more appropriate. For example, training manuals and course notes may form part of the documentation required to train users on the use of a particular product, but more effective results might be obtained by producing a training video, or by combining hands-on experience with some form of audiovisual instruction. Chapter 8 discusses other forms of technical communication is more detail, including multimedia.

Before deciding upon the most appropriate form of technical publication, one should consider the following points:

(a) decide what information the user needs;

(b) decide what kind of information is needed;

(c) decide in what environment the documentation will be used;

(d) if more than one type of information is needed, decide whether it is to be provided in more than one document or combined into one;

(e) find out what resources, timescale and budget is available for the project.

You should then draw up a suitable documentation plan. Unless it is clear what kind of information is to be conveyed and what type of document is to be produced, it may be beneficial to draw up a table indicating the types of information that may be required at different levels of use. An example is shown in Table 3.1 below.

Table 3.1 User needs and documentation types

	Novice user	Advanced user	Service personnel
User guides	✓	✓	
Reference manuals	✓	✓	✓
Support documentation		✓	✓
Overhaul/service documentation			✓
Troubleshooting guides		✓	✓
Training material	✓		

The actual classification will depend on the nature of the technical publication and the subject matter. For example, in some cases, it may be best that the novice user is not given information concerning problem solving or troubleshooting because they may be not have sufficient knowledge to understand the remedial action required. In such cases, it may be better to reserve such information for advanced users and service personnel only as implied in the table above.

Economical and other practical factors can dictate what kind of publication is suitable. Since different types of publication benefit from different approaches to design, it is probably better if they can be produced as separate publications, but often it is necessary to furnish users with a single manual that covers more than one kind of documentation. It may be cheaper to produce in that way and more convenient for distribution. Timescales, resources and costs will all play a role in deciding on the nature of the publication to be provided.

USER GUIDES

User guides are documents aimed at helping the user of some equipment or software application to fulfil tasks for which the system is intended. It may be a document that covers all aspects of use, from first time use to day-to-day operation and periodic maintenance. On the other hand, it may be required to supplement the use of a system by offering advice on how to deal with special circumstances.

The operation of equipment or software may require varying degrees of skill on the part of the user. The amount of training a user is expected to have undergone will therefore determine the level at which the documentation is to be aimed. Not all documentation needs to be addressed to the layperson, even if it is a user guide. There may be valid reasons to expect that the user has undergone a certain amount of instruction before needing to refer to the documentation. Indeed, there are some systems where this is essential, and the user documentation cannot encompass all the information that user will need. As discussed in Chapter 1, the author must assess the readership level before being able to judge the level of content for a user guide accurately.

As an example of the kind of information a user guide may contain, the following is a list of subjects which would be most likely to appear:

(a) a functional description of the system;

(b) guidance on the installation or set up the system;

(c) starting up or first time use instructions;

(d) general instructions on operation or use of the system;

(e) detailed instructions on use of one or more aspects of the system;

(f) task-oriented information advising the user how to achieve a particular result using the system or how to apply the system in a certain set of circumstances;

(g) maintenance of the system if not covered by a separate document;

(h) troubleshooting information to cover possible problems during use of the system, again, if not covered by separate document.

User guides are publications that are most often hybrid manuals that include all kinds of technical publication types in one volume. For example, the documentation may contain some kind of tutorial instruction which helps a novice user to learn about certain aspects of the system before using it in everyday situations and good user guides can often form the basis of course notes for training users in the use of the product. However, in trying to accommodate a whole variety of topics at different levels, they may not be a suitable substitute for proper training documentation and will probably fall short of being adequate for service and maintenance manuals.

One of the most important aspects of documenting user guides is being able to assess the amount of guidance the user requires. Many user guides cover every part of the system concerned, but not all such information may need to be documented. In software documentation, user manuals may be split between a guide on the general operation of the software, perhaps organised according to tasks the user may wish to carry out; with a separate comprehensive reference guide that covers each feature of the system in more detail. A similar approach may be may for equipment documentation. An operator's manual for a piece of equipment may incorporate one or more sections on performing certain functions, while other parts of the user guide explain each of the equipment's functions in detail. The reference section may not be required at all if the purpose for which the documentation is intended is simply to offer advice to the user on the best way of achieving a particular result. Where the user guide also covers maintenance and servicing, it is important to relate the amount of information provided to the level of maintenance the user can be expected to achieve.

The information required to install and set up a particular system may only be referred to once by the user. If the user guide is likely to be extensive in other areas, the author may consider providing the installation information in a separate document, leaving the most important and regularly used documentation readily accessible in its own volume.

REFERENCE MANUALS

Reference manuals can often overlap with user guides. The distinction between the two is that user guides tend to approach the information that users need by organising it in a task-related fashion. In other words, the user looks in the manual for guidance on how to achieve a particular result, and which may require the use of several features of a product or carrying out a variety of tasks in a particular order.

Reference manuals tend to be organised according to the features or specification of a product. For example, in the case of a complex piece of equipment, the reference manual may describe each feature in turn, regardless of the application of the feature. Similarly, in software user documentation, a comprehensive reference manual may described each feature, menu item or option available to the user in either the order in which they be found or in alphabetical order.

SUPPORT DOCUMENTATION

Support documentation can cover a variety of publications intended to supplement standard user guides and reference manuals. They may provide additional task-related information not otherwise covered in the user manual to help users solve particular problems or achieve particular results. They may be used by support staff where the documentation includes additional information about a product not normally required by the user in everyday operation of a product and therefore not covered in the user guide, but which may be occasionally required in special circumstances.

Support staff themselves may require documentation which is a combination of a user or reference manual, a troubleshooting guide and a service manual. The degree of detail give in each case required may be less than would otherwise be provided by separate documents covering these aspects, but organised in a way which makes the manual suitable for first-line support of the product. For example, it may be helpful for a telephone support person to have access to a wiring diagram to help offer guidance to a user in difficulty. The diagram would not be required by the user on a day-to-day basis and is therefore not included in the user's guide. On the other hand, the support technician may not require the level of detail associated with the wiring diagram that might be included in an overhaul or service manual.

In the software industry, it is commonplace for support fact sheets, offering help on specific subjects, to be made available to users. These supplement the user guide by offering help in areas where the user may require up-to-date information, or tips on how to overcome commonly encountered problems that were not apparent to the authors of the user manual at the time it was published.

OVERHAUL AND SERVICE DOCUMENTATION

The deepest level of technical information and engineering procedures are generally found in repair, servicing and overhaul manuals. Complex equipment may need to undergo major overhaul after a certain amount of time service. After a major overhaul, the equipment may need to be tested, perhaps under similar conditions to those employed at the time the equipment was manufactured. Services and overhauls are almost exclusively carried out by highly skilled and qualified personnel. The level of technical content for such documentation is therefore high and the author of such documentation is likely to be someone with a similar level of knowledge or with experience of equipment maintenance.

The documentation may need to cover particular engineering processes, such as wiring, soldering, milling, welding, tuning, etc. There will almost certainly need to be technical drawings covering circuit layouts, exploded parts diagrams, engineering and design drawings, etc. Details specifications will be included and extensive parts lists. Typically, the documentation will need to cover some or all of the following:

(a) a comprehensive description and specification;

(b) technical and engineering diagrams of the system or equipment;

(c) procedure for dismantling;

(d) engineering procedures applicable to the overhaul or service;

(e) fault diagnostics procedures;

(f) maintenance schedules;

(g) instructions for testing serviced equipment;

(h) reassembly instructions;

(i) detailed parts list.

Servicing may be carried out at different levels. To ensure that a particular piece of equipment remains serviceable, it may be sufficient to produce instructions on replacing components rather than repairing. The faulty component may then be passed down to a second line of servicing for inspection and repair. Where a component may be unserviceable, it may be appropriate to include instructions for suitable disposal. This is particularly important if the component is hazardous, such as a battery, chemical container, pressurised container, or non-biodegradable product. Other safety aspects concerned with service and overhaul, such as special handling procedures, must also be covered in the documentation.

TROUBLESHOOTING GUIDES

These types of publication are designed to enable a user or service engineer trace a particular problem and find a remedy. They need to be organised carefully to be effective and this will depend upon the product concerned and what information is given to a user when something goes wrong. If the user is operating a printer, for example, and an error code is displayed on an LCD screen when something goes wrong, the user will need to be able to find that error code in the troubleshooting guide quickly. In such circumstances, it will be beneficial to organise the guide in error code order. If the same printer suffers a paper jam with no error indicator on the equipment, the user will need to find out how to clear the jam in the troubleshooting guide. In this case, it may be appropriate to find the solution in a section entitled 'Paper Handling Problems'. If the guide is well indexed, the user may be able to locate paper jams in the index first, then look up the solution in the appropriate place.

The remedial action covered in a troubleshooting guide must be appropriate to the skills and knowledge of the user. If a problem occurs that is beyond the capabilities of the ordinary user to fix, the guide should refer the user to the appropriate service authority. It is important that where a user is expected to carry out some kind of remedial action, there is no ambiguity over the nature of the problem itself. For example, one problem may arise as a result of several different conditions each of which require different remedial action. If the user cannot determine the cause of the problem accurately, other problem may result because the wrong remedial action was taken.

REPORTS

Although technical reports are generally much smaller scale projects compared to the other publications described in this chapter, they still require structuring and planning with as much care. If a report is intended to inform non-technical persons, it is important to ensure the text uses laypersons terms as far as possible. It is preferable to no have to attach glossaries and it is best not to introduce special document conventions is what will usually be a relatively small document. If the report addresses other technicians, then its is acceptable to use whatever terms, symbols and other conventions will be obvious.

TRAINING MATERIAL

Many technical authors utilise the skills acquired in writing user guides and maintenance documentation to produce training material which may relate to the same subject or project. However, the author may be unaware that there are quite distinct skills required in producing good training course materials. Clearly there is some common ground in that all documentation is intended to 'teach' the reader. The objective in all cases is to inform the reader about facts not already known or remembered. The organisation of materials designed to accompany a training course, lecture, seminar or other instructional activity is critical to its effectiveness. Courseware, as it is known, is better prepared by trainers who have been taught technical authorship. They will usually be better qualified to identify the most important aspects of the subject at various levels of user tuition. Where the trainer is not qualified for technical writing, the technical author employed to write the training material must liaise with the training staff to ensure that the material fits in with the structure of the training course itself.

The amount of detail required in the training material will clearly depend upon the subject involved and the way the training course is run. It may be sufficient to provide notes for delegates highlight key points which may be used as an aid during the course itself. The course delegates may given copies of material used during the course, such as overhead slides, to save them having to make notes while the trainer is speaking.

Training materials for the trainer are of great importance too. A technical author may be required to prepare documentation for the trainer based on existing information prepared for user guides. It is important that the author works under the direction of the trainer. The trainer will know what aspects of a particular subject need to be put across and is best placed to guide the author on the level of detail required at each stage of the course. The technical publications department may be employed to produce the courseware, including designing and preparing overhead transparencies and other audiovisual elements. This requires may involve the skills of graphic artists and technical illustrators.

Some user documentation includes tutorial information. This is usually a section in the user manual itself or an accompanying document that provides a step-by-step guide on how to achieve a particular result, based on an example application of the product. The tutorial will probably begin with a simple task to help the user become familiar with basic operation, then provide more complex levels of example at each stage. In some ways, the considerations for choosing the appropriate content for a tutorial is similar to that of courseware, except that it is usually expected that the user will undertake the tutorial without guidance from anyone else. It must therefore be 'foolproof', as the user will not have a trainer on hand to overcome difficulties and queries.

UPGRADES AND REVISIONS

All technical documentation that is subject to upgrading and amendment must be designed to take into account the means by which the revisions will be supplied to the user. If the documentation is likely to require frequent revisions, the production method needs to enable the original document to be updated easily. Typically, this is done by supplying the documentation in a ring-bound format which facilitates the addition and replacement of pages by the user. If occasional updates are required only, there may be a case for providing supplementary documentation which is self-contained, but is designed only to be read in conjunction with the original document. If revisions are likely to be very extensive, it may be worth considering issuing a new edition to completely replace the original document. In all cases, it is important that the user can make use of the revision documentation conveniently. If supplementary documentation is supplied in the form of a separate, self-contained document, the user must be able to relate the content with that of the original one. The user will be frustrated if there is a poor correlation between the revision documentation and corresponding information in the original.

The author should remember that revision documentation supplied as a separate addendum can be separated from the original manual or simply lost. If the revisions are likely to impart information which is critical to the continued operation of the product or system that may have been upgraded, the author must consider the consequences of supplying the revision documentation in a particular way. The amount of information supplied in self-contained addendum will depend on the nature of the revision. It is fair to assume that the reader will have knowledge of the conventions used in the original documentation and will be familiar with subjects that are covered in the original. However, it can be frustrating for a user to have to refer to two separate documents in order to obtain the information needed, when the revision documentation could have provided all the information in one place.

If revisions are supplied as replacement or additional pages to a loose-leaf document, the user must be given clear instructions on how to upgrade the original document. It will be helpful to summarise the changes made, so a revision history can be maintained by the user and he or she can check that all the relevant pages have been added or replaced. It may be helpful to add a date or version reference to the revision pages. In this way, it is possible to determine, at any given point in time, whether the user has the latest version of the documentation. Like supplementary documentation, revision pages can get lost, particularly if the user fails to update the loose-leaf original as expected. Therefore the author must consider whether the addenda should be accompanied by warnings or disclaimers to impress on the user the importance of carrying out the update.

SUMMARY

1. The type of publication chosen for a particular purpose must be carefully planned and take into account the user's needs and what type of information must be included.

2. Although most technical documentation is paper-based, the author must consider whether printed matter is the most suitable means of conveying the subject matter.

3. User guides provide a variety of information to assist in the operation of a system or piece of equipment. They may include information at all levels, from first time use through to troubleshooting and ongoing maintenance, but the author needs to assess whether some information is best supplied as separate publications. User guides are best organised in a task-related way.

4. Reference manuals are designed to provide comprehensive information about every feature of a product or system. They may be organised in the order in which features will be used or operations carried out, or some other convenient way of referencing, such as alphabetical order of subject.

5. Support documentation may be supplied as supplementary information to assist users in achieving particular results, or may be special documentation used by support staff to offer user's guidance on particular aspects of the subject when queries arise.

6. Overhaul and service documentation involves the deepest level of technical writing and may include detailed engineering procedures, technical illustration, test procedures, disassembly and reassembly instructions, etc.

7. Troubleshooting guides must be carefully organised to ensure that solutions to problems can be found quickly. The level of remedial instruction must be appropriate to the skills and capabilities of the reader.

8. Technical reports, although usually brief documents, must be written at precisely the right level of reader knowledge, as there is little scope for glossaries of terms and other supplementary material.

9. The preparation of training material requires special skills and technical authors must operate under the guidance of the training staff.

10. Technical documentation should be designed to take into account the mechanism that will be used for future revisions. The way in which documentation revisions are supplied depends upon the frequency with which updates will be required and the extend of material revised.

FURTHER READING

Bly, R. W., Blake, G. *Technical Writing — Structure, Standards and Style* (McGraw-Hill, 1982).

Hartley, J. *Designing Instructional Text* (Kogan Page, 1985).

4 Documentation Design

Successful presentation of technical documentation relies on good knowledge of the basic principles of layout and design. The purpose must be to give a visual form to the work by arranging the elements of typography, illustration and space in such a way as to communicate the content of the text in the most effective way.

By concentrating some effort in this direction, many technical documents would be received with a far warmer reception than they generally do. For those non-technical readers who have to read technical material in order to understand the working of some system or product, for example, the mere thought of reading material of a technical nature can be ominous. Therefore, if is the duty of the publisher of such works to provide some encouragement. The layout and design should aim to do at least the following for the reader:

(a) invite the 'opening' of the documentation in the first place;

(b) give the impression that the contents are easy to follow;

(c) encourage reading of the text; and,

(d) help the reader find what is wanted.

If the design and layout of the document fails to do the above, then no matter how accurate or well written the text, the document will not serve its fundamental purpose of communication.

Finding the right design for any particular document is not easy. There is no one solution or recommended design that can possibly suit every type of documentation. Documentation must be designed to make it suitable for conveying its message to the reader, having taken into account a number of factors. Designs suitable for reference manuals are not generally appropriate for tutorials or quick-reference guides. The design for a reference manual to accompany a piece of electrical hardware is not necessarily appropriate for the reference manual of a programming language; and a design for documentation of a word processing application is not necessarily suitable for accounting software.

It has become commonplace, particularly with user manuals for application software, to copy trends in design set by major players in the market. Indeed, many designs are driven by a need to satisfy corporate identity or to follow a house style and not the need to satisfy the objective of the documentation, which is to communicate its message to the reader in the most effective way possible. If one piece or a set of documentation needs to convey different types of information for the same product or service, then the design should take into account the best way of conveying each element. If this means a diverse range of styles would appear in one document, then consideration should be given to breaking down the document into separate parts. For example, some software user manuals include general background reading, a tutorial, quick-reference data, a comprehensive reference guide and troubleshooting information all in the same publication. In applying a house style across all such sections, no one section is given the design best suited for its purpose. It would be better, in such circumstances, to produce separate publications for each section. It may even be of more benefit to the user if the tutorial is separated from the reference section and quick-reference guide.

DESIGN AND LAYOUT

Design and layout cannot be simply a matter of guesswork if it is to work well. With the widespread use of desktop publishing and sophisticated word-processing software, the tools for layout are in the hands of many who have not been trained in the principles of typography and design. The best layouts may be produced from a combination of skills which require the technical knowledge of graphic artists, typographers, photographers, art editors and the like. While it may be impractical to have access to such resources every time a document needs to be designed, one should never underestimate the specialist nature of this work.

To help the novice designer, some desktop publishing systems offer template designs for various types of document. These may provide a useful starting point, particularly if the user is stuck for ideas. In the absence of any particular house style to follow, one can benefit from looking at examples of similar technical documentation. When evaluating other publications, try to work out what effect the layout has on the overall use and appearance of the documentation. Some examples will be more 'comfortable' to the eye than others.

Page Size

The appropriate choice of page size and orientation will depend upon the method of production and what is practical for the type of information contained in the publication. If the document is to be photocopied, then it will be more convenient if the document is

A4. If it is to be litho printed, one has much more choice of page size available. Refer to the information given under the Finishing and Packaging heading, later in this chapter.

Page Layout

This is partly dependent upon the choice of paper size mentioned above. However, it is always preferable to choose simple layouts that can be prepared easily and without the need for specialist graphical skills. The layout will vary according to what type of information needs to be presented on the page. It is helpful to have an overall structure to the page layout which is uniform throughout the document as far as possible. There may be occasions when certain tabular information, illustrations, lists or other special details needs to be presented in a way which varies from most of the rest of the document. However, one should begin with an overall grid in which most of the page elements can be located. The following points should be considered:

(a) number of columns to be used for the text;

(b) margins and indents required;

(c) levels of text and headings required;

(d) positioning of tables and figures.

Try to imagining your type or text as blocks on the overall page. A variety of effects can be achieved by arranging blocks to create the overall usage of space on the page. Increasing white space around the text blocks may be employed to achieve the following:

— providing space for notes;

— creating an impression of space to make the document seem like lighter reading;

— add distinction to the document.

By reducing the white space around text blocks, the following may be achieved:

— allowing more text to be printed on the pages, resulting in fewer pages in the whole document;

— improving the appearance if the typesize is large or generously spaced.

Text Columns

Figure 4.1 shows example arrangements of text blocks on facing double pages. Whatever arrangement is chosen, a balance should be created between the type and space that best suits the purpose of the documentation and the type of information to be conveyed. Simply choosing a design for its aesthetic qualities is not recommended. Where multiple column layouts are chosen, it important to choose correct type size and character spacing for the width and depth of columns used (see also information given under the heading Typography, later in this chapter). Generally, no more than two columns should be necessary, unless the orientation of the page is to be changed from portrait to landscape.

54   TECHNICAL DOCUMENTATION

Figure 4.1 Alternative arrangements for text layouts on facing double pages.

The 'gutter' or spacing between columns needs to be chosen carefully as should the indent of the text blocks from the top, bottom, left and right margins of the page. These may be affected by the method of binding used. If a document is to be ring-bound, it will be important to leave sufficient margin for the holes to be punched. If pages are to be printed both sides, the indents may need to be different for left- and right-hand pages. Most modern word processor and desktop publishing systems can control the spacing and positioning of multiple column text blocks automatically and usually offer default indents and gutter measurements which are adequate for most purposes.

Spacing may be needed beside text blocks for the user to write notes. This may be applicable for training course material and educational documents.

Other Page Elements

Consideration will need to be given to allow room for running headers and footers that accommodate the page number, chapter number and title, etc. If footnotes are to appear, space will need to be provided for those also. Decorative lines may be used to enhance the appearance of the page. A line across the bottom of a page can help to give the appearance of a complete page, even if there is only a little text at the top. It is commonplace to use lines to separate running headers and footers from the main text area on the page. The weight of the lines used should be in proportion to the page size and typefaces used. If the lines are too heavy, they will be distracting. Printing in another colour can help tone down lines (see also information under the heading of Colour later in this chapter).

The placing of illustrations, tables and various figures must be defined. It may be appropriate to enclose figures in boxes, in which case it will be more important to ensure that the boxes align with text blocks and other page elements in order to keep the page looking tidy. The amount of space afforded to illustrations will depend on the type of illustration to be included and the page size. It can be helpful to draw a rough page layout in the form of a grid to plan the position of all page elements. Figure 4.2 shows an example page grid as used by a British Standard A4 document. Figure 4.3 shows a similar grid for an A5 page. Figure 4.4 shows an example of an A5 page with text. These examples have been reproduced by courtesy of the British Standards Institute from BS7649: 1993.

You may wish to use a different arrangement of page elements for chapter and section beginnings as in this book, where the position at which text begins on the first page of a chapter is different to the subsequent pages. This serves to break up the monotony of the document and may help a user to identify the beginnings and ending of chapters and sections more easily.

The degree of control that may be exercised over the page elements will depend upon the system used for production of master pages or camera-ready artwork. The more comprehensive the word processing and desktop publishing system used, the more typographic freedom there will be available.

56 TECHNICAL DOCUMENTATION

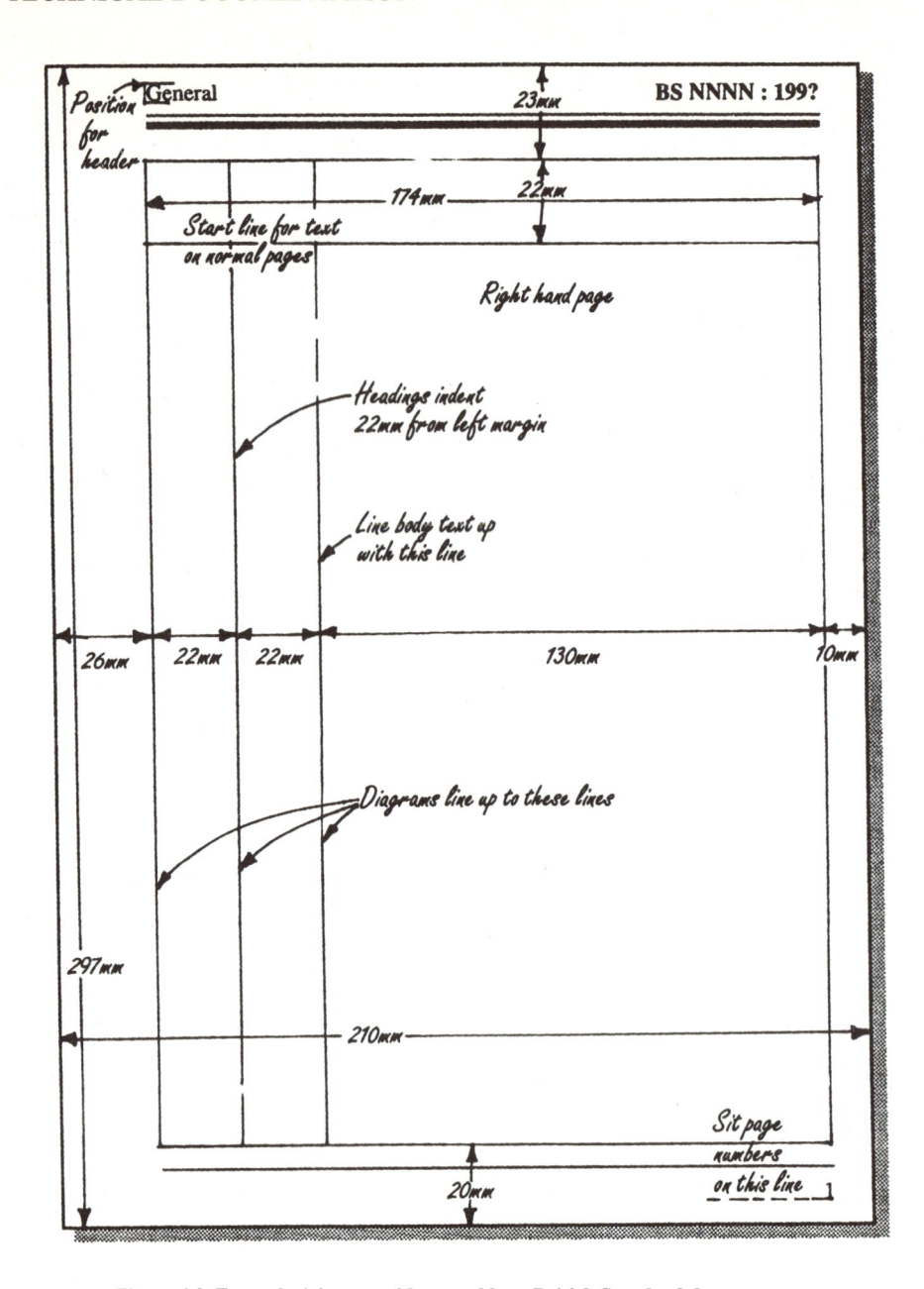

Figure 4.2 Example A4 page grid as used by a British Standard document.

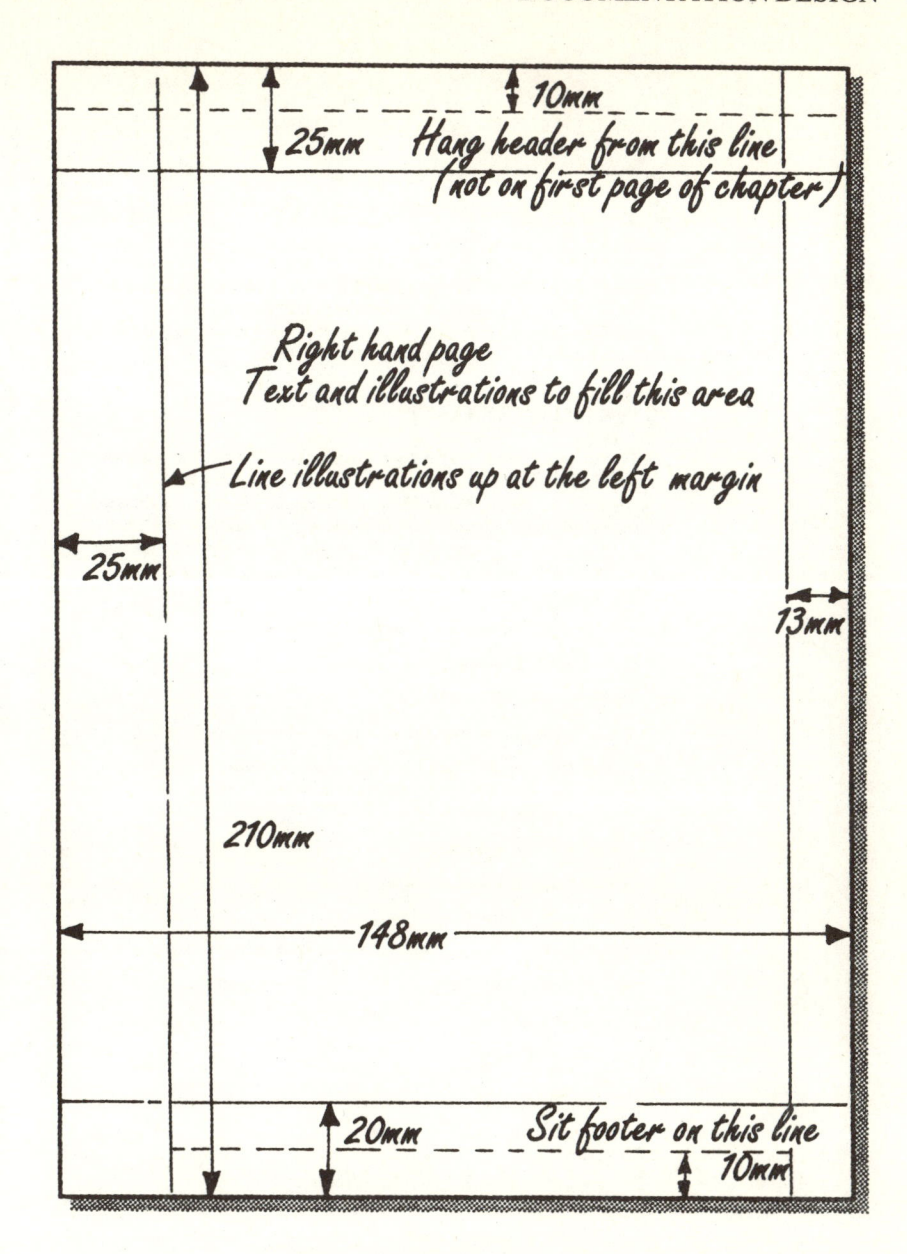

Figure 4.3 Example A5 page grid.

What is EAS? Chapter 1

This chapter introduces you to your Example Application System (EAS). It outlines the tasks you can carry out using EAS, it explains what a menu system is and how it works, and it describes the equipment you use when you work with EAS.

What can I do with EAS?

EAS saves you time and effort when you carry out the part of your job that involves record keeping and reporting. It performs calculations for you and it reduces the chances of making a mistake when you record information. If you use a communications link with EAS you are not dependent on the postal service because you can send your reports to head office electronically, using the telephone network.

EAS is divided into two main parts:

- The Weekly Statement of Business
- Staff Time and Attendance

Your EAS system may consist of only one of these parts.

You use EAS to record information on either a daily or weekly basis as required. Examples of the sorts of information you may need to record are given below:

- Income for drinks for a session
- Income for food for a session
- Petty cash payments
- Receipts for expenses such as window cleaning or laundering
- Hours worked by each member of staff in a session
- Staff holiday hours
- Amounts banked

Figure 4.4 Example A5 page layout.

TYPOGRAPHY

Most reference manuals and books are typeset in what is called a 'book face'. This is typeface which is easy to read in long passages. If a large amount of text were to be set in a fancy and decorative style, it may look pleasant at a glance, but if it makes the reading of it difficult and slow, it will not be appropriate for the purpose. Finding an *appropriate* design is part of what make good typography. The more one learns about design, the more knowledge and relevant information one will have to apply to the next publication.

Typefaces

Typefaces fall into one of two main categories: *serif* and *sans serif*. Until the 1800s, nearly all typefaces were serif faces. A serif typeface is one in which each character has a short stroke at right angles or oblique to the main stroke or arm of the character (see Figure 4.5).

N

Figure 4.5 A serif character.

Serif characters come is a huge range of designs. Probably one of the most widely used and well known being 'Times'. It was introduced by Stanley Morison, a consultant and typographic advisor to the Monotype Corporation, the Cambridge University Press and *The Times* newspaper of London. Morison was a dominant force in typography and printing from the 1920s and his design of Times Roman was probably one of the most successful typeface designs of the 20th century. As mentioned earlier, serif typefaces included most classes of typeface used for bookwork and continuous setting, where readability is important over long passages. It is generally considered unwise to mix different serif faces together in the same publication and is usually pointless, as many of them are quite similar in appearance.

Sans serif characters, that is, those that do not have serifs, did not appear until quite late in printing history, the most well known being Helvetica, Univers and Gill. These tend to be most suitable for headings, captions and display work.

Why are certain typefaces chosen in preference to others? The answer to this question will always be a subjective one. Decisions about the right typeface for a particular job tend to come down to a mixture of common sense and personal preference. As far as technical documentation is concerned, decisions about which typeface to use are less important than they are for other types of publication, such as sales and marketing literature. It is most important that the documentation is easy to read and other design aspects such as typesize, spacing, line width, etc. are probably far more important to get right than the actual typeface used. There may be a case for selecting a particular typeface because it follows a corporate style, but otherwise it is simply down to what suits one's taste best; there is certainly no lack of variety available.

Typesize

In typesetting, the size of the typeface is quantified by its *point size*. The point (which is used with the abbreviation 'pt') is the basic unit of measurement in typography and most other dimensions used in printing are derived from this one measurement. The point is actually $1/72$ of an inch and is used for the measurement of spacing as well as typefaces. It is considered that sizes between 9 and 12 point are generally suitable for the main text in a technical publication, depending upon the page size, line width and typeface chosen. Sans serif typefaces, generally chosen for headings tend to be more prominent that serif faces, even if the same point size, but headings generally need to be in a larger size and different weights and sizes of type should be chosen for different levels of headings in a suitable hierarchy.

Typestyles

Emphasis can be given to type by changing its typestyle from normal (otherwise called roman) to a style such as bold, italic or underlined. Care should be taken to ensure that the emphasis given to a particular piece of text is appropriate to the purpose.

Generally, bold type always attracts attention. Italic type appears 'lighter' than normal type, but still stands out. Although it is not unusual to see underlined headings in newspapers, underlining is not generally considered necessary for emphasis and can be unattractive, particularly when used with headings; there a far more effective means of providing emphasis using different weights and sizes. The following is a list of recommended uses for bold or italic text. Use bold for:

— chapter, section and paragraph headings;

— figure captions;

— illustration and table titles;

— important text, messages and warnings.

Use italics for:

— emphasis of special words, phrases and terms;

— variables.

Different typefaces may also be chosen to provide emphasis, such as a monospaced typeface for system messages in software documentation. Figure 4.6 shows an example of the use of text attributes. This example has been reproduced courtesy of the British Standards Institute and the page shown is reproduced by permission from ICL.

DOCUMENTATION DESIGN 61

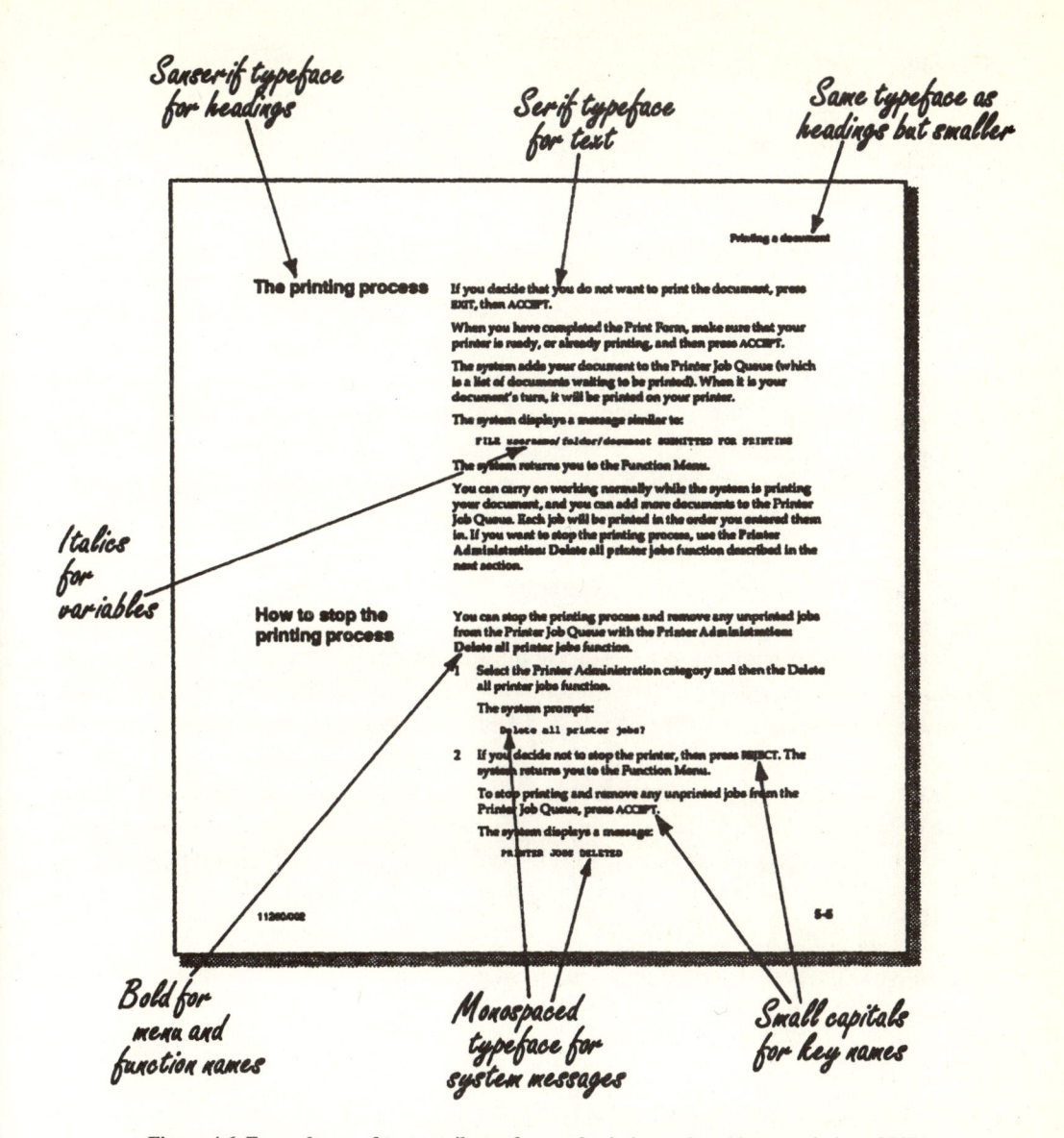

Figure 4.6 Example use of text attributes for emphasis (reproduced by permission of ICL).

Spacing

Spacing in typography affects many qualities of a publication. There are many spacing considerations that can determine whether a section of text looks right on paper. Each of the spacing attributes may be altered in isolation or together to affect the appearance of the publication. There are four main categories relating to spacing type and most word processing, desktop publishing and typesetting systems cater for them in one way or another, through the degree of control afforded to each may vary. These categories are:

(a) character spacing;

(b) word spacing;

(c) line spacing;

(d) spacing around the text block (that is, margins, gutters, etc.).

All these affect the appearance of the typeface when they are changed. The spacing between characters comes under a special terms, known as *kerning*; a method of space adjustment used particularly with large type. Control over character and word spacing is something that is peculiar to typesetting unlike a traditional typewriter. A typewriter uses an even amount of space for each letter or character, thus the letter 'i' takes up the same width on the paper as the letter 'm'. A space also occupies the same width as any character. Consequently, the space between any typed word is constant. This is known as non-proportional spacing. In typesetting, the spacing between words and characters can be changed. To begin with, characters of physically different size can occupy different widths as can be seen in the setting of this book. This is known as using proportional type. Spacing between words can also be varied in order to 'spread' the words on a line to 'justify' or align it on both left and right margins; again, as in this book.

Most word processors and typesetting systems can offer both proportional and non-proportional typefaces.

COLOUR

The use of colour in technical documentation is still relatively rare given the vast quantities of publications produced. Where additional colour is employed, it is generally confined to a 'spot' second colour, where certain elements of the page design or key text and headings are picked out in a colour or shade other than black. It is also generally confined to publications that are produced in large numbers, where the additional cost of colour can be justified by the low unit costs that result from high quantity print runs or from very low print runs where a colour inkjet or laser printer can be used to print the actual number of copies required.

The use of at least one additional colour will enhance the appearance of any document if used wisely. The affect of colour on design of the page is significant and can certainly change the priority of particular elements of text if they are printed in a different colour.

The choice of colour used will also have an effect on the way the documentation is read. Using soft tones of green and blue can add interest to a page without necessarily changing the emphasis of any particular element. Using a strong red in the text can spell 'danger' and may only be suitable for warnings, for example, to highlight health and safety aspects, though care must be exercised when relying on colour to convey meaning since it can easily be overlooked or misinterpreted.

FINISHING AND PACKAGING

Finishing and packaging of technical documentation should be decided upon after consideration of the following:

(a) the budget available;

(b) the quantity to be produced;

(c) the frequency and methods employed of revisions and updates;

(d) marketing considerations.

In engineering, aviation, defence and other such areas of technology, A4 remains the most common format for technical publications. Many technical documents are produced in short print runs. Many are prepared on laser printers and reproduced by photocopying, so A4 is a practical format in both these circumstances. For computer software documentation however, much of which is published in the USA, format varies widely. There production numbers are usually considerably higher than those in other areas of technology.

With almost 30% of technical documentation being produced by photocopying, there are limitations to how the presentation of the document can be improved by the use of other materials. There is a range of qualities of papers suitable for photocopying offering various degrees of opaqueness and whiteness. Although a heavier paper can enhance the quality feel of the document, again, photocopying usually limits this to around 90 gsm. Smoother paper may be more attractive, but too smooth a paper does not hold toner well and other attributes such as moisture content can also affect reproduction quality.

Those documents that are printed through an offset process can vary in quality depending upon the paper used. There are many papers to choose from and different paper surfaces will suit some types of printing better than others.

How a publication is going to be finished and presented may affect the design and layout characteristics. A report that is going to be comb-bound, or ring-bound, for example, may require a layout that provides extra space in the left-hand margin, to avoid holes being punched through the text.

The finish of documents is even more important when litho printing is being used. In the publication of a book or manual, where both sides of the paper are printed on, the gutter between the two pages needs to be greater at the centre fold than at the outside margins of the page. Many desktop publishing systems will produce this facing and double-sided page effect automatically by selecting the option as part of the page layout attributes. If this extra space is not provided, then when the book is bound, the text near the centre of the page will be more difficult to see, and this problem is accentuated by the process of perfect binding (see description later), in which the book's pages do not lie flat as easily as they would if comb- or ring-bound.

Paper Sizes

The finish of a publication may also dictate the paper size used. Some processes are cheaper if standard sizes are used for binding equipment and accessories. For example, ring binders are available in standard sizes for A5 and A4, whereas other sizes may have to be assembled specially to order. While this is not in itself a problem, the economics of the print job have to be considered too. Obviously, any system of binding which deviates from standard facilities and practices is going to be more expensive as a result. As a guide to standard paper sizes, the following table has been included for reference purposes.

Table 4.1 'A' series paper sizes

Size	Inches	Millimetres
A0	33.11 × 46.81	841 × 1189
A1	23.39 × 33.11	594 × 841
A2	16.54 × 23.39	420 × 594
A3	11.69 × 16.54	297 × 420
A4	8.27 × 11.69	210 × 297
A5	5.83 × 8.27	148 × 210
A6	4.13 × 5.83	105 × 148
A7	2.91 × 4.13	74 × 105
A8	2.05 × 2.91	52 × 74
A9	1.46 × 2.05	37 × 52
A10	1.02 × 1.46	26 × 37

This system for sizing paper was established in Germany in 1922 and is sometimes referred to as DIN A size. Each size is arrived at by halving the size immediately above it, so it is simple for scaling purposes, since they all have the same geometric shape. The 'A' paper sizes always refer to the trimmed sheet size, untrimmed sizes are referred to as RA or SRA. Another series of paper sizes used for those in between the various 'A' sizes is the 'B' series, but this is relatively uncommon.

Apart from standard paper sizes, it is useful to known the standard book sizes for

DOCUMENTATION DESIGN 65

book and manual print finishing purposes. These are shown for the United Kingdom (UK) in Table 4.2 and the United States (US) in Table 4.3. Note that the UK book sizes show the names of the various formats which might be referred to by a printer.

Table 4.2 Standard US book sizes

millimetres	inches
140 × 216	5.500 × 8.500
127 × 187	5.000 × 7.375
140 × 210	5.500 × 8.250
156 × 235	6.125 × 9.250
136 × 203	5.375 × 8.000
143 × 213	5.625 × 8.375

Table 4.3 Standard UK book sizes

Quarto

name	millimetres	inches
Crown	246 × 189	9.69 × 7.44
Large crown	258 × 201	10.16 × 7.91
Demy	276 × 219	10.87 × 8.62
Royal	312 × 237	12.28 × 9.33

Octavo

name	millimetres	inches
Crown	186 × 123	7.32 × 4.84
Large crown	198 × 129	7.80 × 5.08
Demy	216 × 138	8.50 × 5.43
Royal	234 × 156	9.21 × 6.14

There are a whole range of papers used for printing or publishing. If printing is done from desktop publishing artwork, the paper used must suit the printing process and the appearance and weight should be considered carefully. The weight of the paper is important where printed material is being sent by post, since a heavy paper can add considerable cost to the mailing bill, especially if the quantity is large. While heavier papers have a higher-quality feel about them, and have less of a 'show through' problem (most apparent on double-sided printing), it can be costly to mail.

The weight of paper is generally measured in gsm or g/m^2 (grams per square metre). This is the weight in grams of a sheet of paper one square metre in size. Paper may also be referred to in pounds (lb) weight. Laser printer paper now commonly available for use with desktop publishing systems is a high-white paper of 100 or 115 gsm weight, while normal photocopying paper is 80 gsm. Most laser printers print more successfully on papers that have a fairly rough surface on which the toner can key. Exceptionally smooth papers or artpapers tend to cause the toner to smudge during the printing process.

Trimming and Binding

As already mentioned, the method of binding to suit the publication and budget of the job may affect the style and design of the artwork required. There are various ways of binding documents and books, which range as much in price as they do in quality. Convenience is another factor to be considered when choosing the appropriate binding. For example, in a document that may require constant updating, a ring-bound method may be the most suitable. Any method of binding which cannot be updated without destroying the existing binding may be very inconvenient for technical information, which is subject to change on a regular or *ad hoc* basis.

Convenience may be important if binding and finishing processes are to be carried out in-house. To produce an A5 document from laser printer A4 paper sheets, trimming or folding facilities may be required. To help with trimming, many desktop publishing systems allow crop marks to be printed on the paper, where the size of the printed page is less than that of the paper size used in the cassette of the printer. Even then, trimming many sheets to size can be cumbersome without the right trimming equipment and many desktop publishing systems print the page in the middle of the paper. For example, an A5 sheet with crop marks is almost always automatically centred on the paper, meaning that the trimming must be done on four sides. An alternative might be to print two A5 sized pages, side by side on A4 paper, so that it can be cut in half to make the A5 sheets — one of the advantages of the 'A' series paper sizes. The printed sheets are collated and folded and trimmed to give the appropriate finished paper size. (Most printed sheets are trimmed after folding.)

The trimming requirements may vary according to the method of binding adopted. If the paper is to be folded and stitched, the pages will need to be trimmed after folding, otherwise the edges of the paper will be misaligned, especially if there are many pages in the publication. It may be necessary to consider leaving larger gutters between pages, where the document is to be folded into a large section and stapled, since the pages at the bottom of the folded stack will need more space between the text than those at the top. Figure 4.7 illustrates various binding methods. Some are briefly described in the following paragraphs.

DOCUMENTATION DESIGN 67

Ring binding

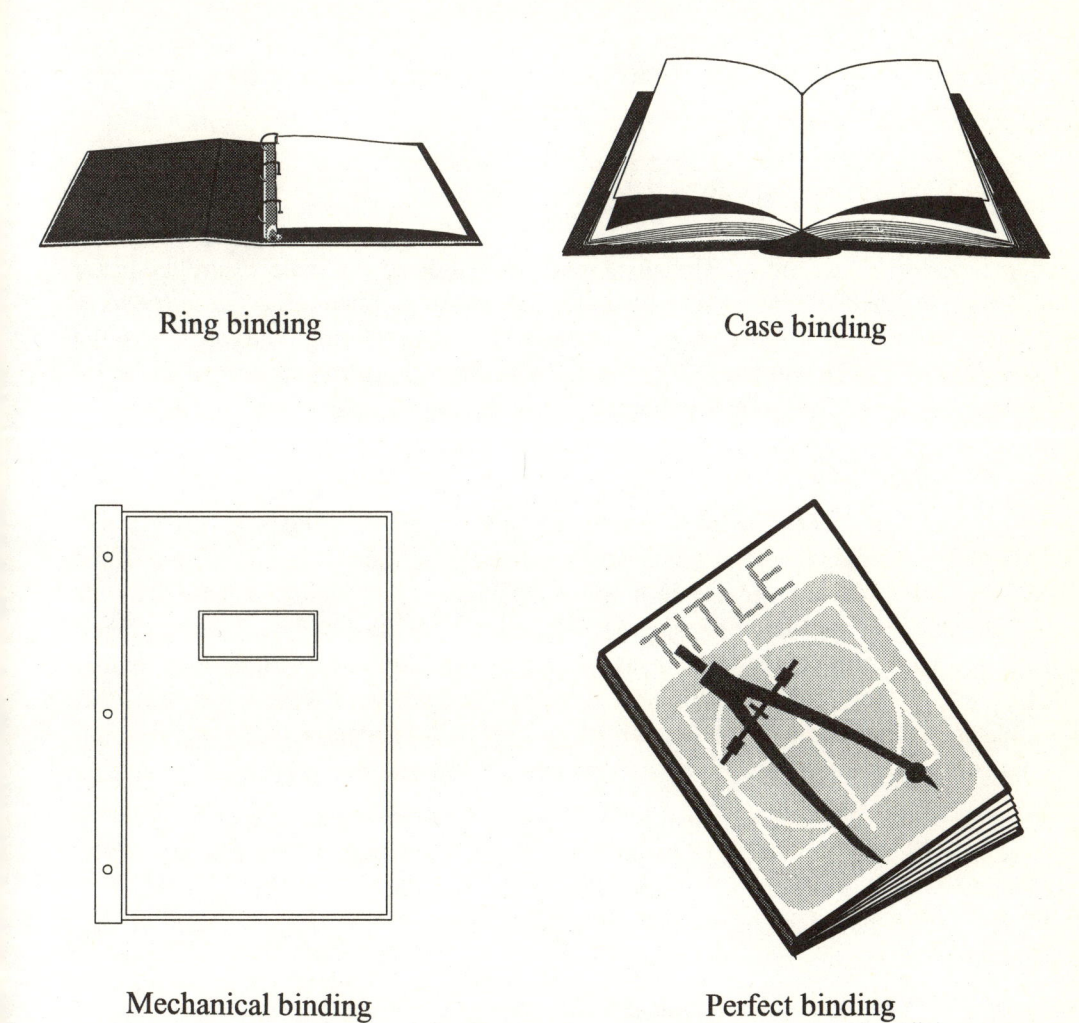

Case binding

Mechanical binding

Perfect binding

Figure 4.7 Examples of binding methods.

Ring Binding

This method is favoured for small publications, training manuals and other technical documentation. It has the advantage of being easy to do in-house, it is relatively cheap, and is suitable for any publication that requires regular updating. Ring binding also enables the publication to lie flat on a surface which makes a document that is required for constant reference easier to use.

Popular sizes of ring binders are A4 and A5. Many organisations have special binders produced with blocking or silk-screen printing, and/or printed inserts encapsulated on the outside to identify the product or the contents of the publication. The number of rings in a ring binder varies. The more rings in the binder, the more secure the paper within it. 'D'-shaped rings are also better for holding large documents, since they stack flat against the binder and do not curve around as with an 'O' ring. The capacity of the binder must be sufficient to cope with the size of the publication, otherwise the pages become damaged. The weight of the paper may affect the capacity too.

Mechanical Binding

This can take several forms from comb-binding to Wire-O binding or plastic gripper. These are all binding methods which can be carried out in-house if the quantities of the publication are not too large. Comb-binding and heat-seal binding are popular methods for presenting technical reports, specifications, technical product details, etc. Both are available from most office suppliers as desktop systems, and heat-sealing is becoming a favourite over comb-binding for high-quality presentations. Note, however, that heat-sealed documents do not lie as flat as comb-bound publications, so larger margins will need to be left at the centre of the pages to compensate for this.

Comb-binding devices are suitable for documents of varying sizes, and the capacity of combs changes according to the size of the job to be bound. The position of the punched holes at the edge of the paper can also be altered; they can be punched much closer to the edge of the paper and still provide a strong binding, compared to ring binding.

Grippers, like heat-seal bound documents, do not allow publications to lie flat, and have a tendency to spring open if the capacity is stretched to its limit. Therefore, give plenty of room in your centre margin for the document to be opened without having to put a strain on the binding.

Wire-O bound documents are usually assembled using the machinery of a print finisher. It is not a cheap method of binding, but is stronger than comb binding, while having the same advantages of opening flat and requiring narrower margins than other forms of binding. Wire-O binding is often used for reference manuals and literature which requires constant use, but may also require the presentation quality which is achieved by this higher-quality binding method.

Perfect Binding

Most paperbacks are perfect bound. This method employs the use of glue to hold folded and gathered sections together with the cover. The back of the folded pages are usually trimmed and roughened to help key the sections to the cover as a complete block, when the adhesive in applied.

The advantage of this method is that it provides a professional look and is a relatively cheap method of binding where large numbers of a publication need to be bound. The disadvantages include a tendency for the sections to become unstuck, particularly if the publication is used regularly, and, therefore this method of binding is not suitable for reference, service and maintenance documentation or training manuals.

Sewn Binding

With this method, after the sections of a publication are gathered and folded, a sewing machine inserts threads through the spine of each section and then uses another thread to sew the sections together, to form a single block. This is then glued to the cover like a perfect bound book.

The advantage over unsewn perfect binding is that the pages are more likely to stay together since they have the added strength of the thread. This method is employed in both paper and hardback bound publications.

Case Binding

Case bound books are nearly always hardback books. The 'end papers' are glued onto the first and last sections, which are glued down to the case. The sections themselves are stitched in the same way as described above, but a linen or paper lining is glued to the spine to help reinforce the joint where the case is applied. Head and tailbands (folded strips of cloth inserted at the top and bottom of the spine beneath the lining) may be used to make the binding look more attractive. The book block can then be left with a flat spine, or may be rounded and backed which gives a firm grip to the sections and helps prevent the middles of sections from dropping forward. Case binding is usually reserved for special quality publishing or where the publication quantity is relatively low. It is rarely used for technical documentation except where prestige is of paramount importance.

Imposition

Having produced page artwork, a printer will need to orientate the pages in such a way so that, for each side of a printed sheet, the correct order will be achieved once the sheets are folded and cut. It is this process which often influences the choice of the number of pages that are included in a publication, since the effective use of the printed sheet area can affect the cost of a print job. It is for this reason, for example, that a manual might be more economically printed if multiples of eight-page segments are used, rather than other combinations.

70   TECHNICAL DOCUMENTATION

The imposition of an eight-page segment might be like this:

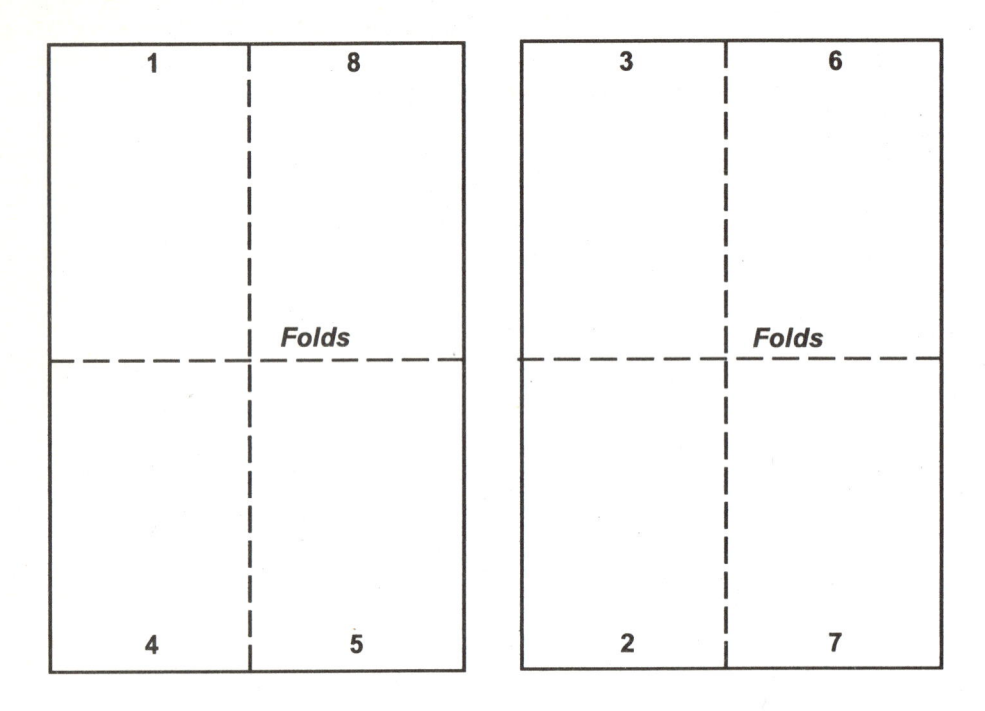

This would result in the following eight-page segment:

Paper is an expensive part of the printing process, so careful planning is important to make the best use of it. When printing long technical documents, the following formula might be useful. To work out the number of sheets of paper that are needed to print a brochure or book (excluding the cover):

$$\frac{\text{No. of books} \times \text{No. of pages in book}}{\text{No. of double-sided pages}} = \text{No. of sheets required.}$$

To work out what number of copies can be produced from a given quantity of paper, the following formula may be used:

$$\frac{\text{No. of sheets} \times \text{No. of double-sided pages}}{\text{No. of pages in book}} = \text{No. of copies.}$$

The size of the page when trimmed (actually referred to as the trimmed page size) will determine what kind of printing machines and finishing will be used on the print job, and, therefore, the cost. A small print job produced on a simple press will be cheaper than if produced on a complex, fast press, since the setup time will be in proportion to the scale of the task.

Using Table 4.4 as a guide, according to the number of pages in the publication, one can see what the maximum trimmed page sizes would be from a press size sheet measuring 1100 × 1600 mm. To maximise on the use of a press, page sizes which are closest to the maximum for the number of pages required should be selected. Apart from the page size and use of colour, the cost is particularly dependent on the print run (the quantity of copies). This also affects the kind of printing machine on which a job will be produced.

When assessing the quantity to be printed, it is always worth asking the printer to quote the run-on rate. This is the cost for an extra number of copies for the same job, if printed at the same time, and is usually given as an average number of pages. For example, the cost for printing 1000 copies with a run-on rate of the cost for an extra 100, 200 or 500 copies. In most cases, the run-on unit cost of a job will be considerably less than for the initial quantity quoted. This is because, while the press is setup and paper is available, the cost of running off 100 or so extra copies will be largely a material one. It might be worth having the extra copies printed if they can be used within a reasonable timescale, which will be cheaper than getting a reprint later.

Types of paper to use

There is a vast range of paper qualities available for various types of printing jobs, and the choice of paper depends on the type of print job, the effect required, the cost of the job and the suitability of the paper to the print process being selected. Some of the more widely used classes of paper are described in the following paragraphs.

(a) Wood-free — paper described as wood-free is not, actually, free of wood; it is still chemically made from wood pulp. Wood-free paper is strong, carries a good whiteness, and is suitable for a wide range of uses where quality and opaqueness is important.

(b) Cartridge — cartridge papers were originally used for making cartridges. They are strong, but not very white. White cartridge has been bleached and is used for drawing paper. Offset cartridge is perfectly even-sided and is used for quality offset printing.

(c) Art paper — artpapers are coated papers. They are usually coated with china clay and calendered (a smooth finish produced by passing paper through polished steel rolls of varying temperatures and pressures) to give a high-quality gloss surface. They are particularly suited to high quality, fine detail work and can easily cope with the reproduction of half tones and colours. Other types of art paper include the matt artpaper which is also coated with china clay, but not given the high polish of artpapers in general. The surface is still very smooth, however, and therefore suitable for high-quality reproduction. Apart from the art papers which are machine coated, there are 'imitation' artpapers which are produced by adding a mineral loading to the wood pulp, and then they are highly calendered. Imitation artpapers are quite good for reproduction of half-tones and line work, but do not have the high quality of the machine-coated papers.

SUMMARY

1. Designs for technical documentation need to suit the information being conveyed. The page size and method of production will affect the choice of design, but page layouts should always be kept simple.

2. The page layout can be planned as a rough grid to determine text blocks, columns, margins, indents and the position of other elements such as decorative lines, illustrations, tables, etc.

3. Typographical aspects of documentation involve choosing appropriate typefaces, typesizes and styles. Attributes that affect the emphasis of type should have significance and be used consistently.

4. Colour may be used to great benefit, but it usually involves additional costs which make its use prohibitive except in cases where large print runs of documents can be justified.

5. Choosing the appropriate finishing and packaging for documentation will depend upon a number of factors, including the method of updating, quantities to be printed, environment in which the documents will be used, etc. Other aspects, such as choice of paper, again depends on the purpose of the documentation and the final quality required.

FURTHER READING

Marlow, A. J. *Good Design for DTP Users* (NCC Blackwell, 1990).

Miles, J. *Design for Desktop Publishing* (Gordon Fraser Gallery Ltd, 1987).

5 Word Processing

There can be few instances now of those involved in the publication of technical documentation not using the facilities of a word processor. By now, word processors are in widespread use in both home and offices and the benefits that they bring to authors are well known. Word processors generally fall into one of two categories; simple word processors designed to handle only text with limited control over the output and those that border on being desktop publishing systems. The distinction between some word processing software and desktop publishing is becoming hard to identify. This chapter discusses the some of the most important features of word processing and does not distinguish between various types of word processing application. Some of the features described here may be just as relevant to a desktop publishing system. Although, as in many other areas of software application, the range of packages available on personal computers is considerable, a few stand out as being universally accepted as setting the standards and are in widespread use throughout the world. These include the likes of Microsoft Word and WordPerfect that are used both in IBM-compatible and Apple computer environments. This brings a benefit in itself; the ability of most word processing applications to be able transfer documentation files between themselves (known as 'importing' and 'exporting') to greater or lesser degrees, so it matters less whether authors are using the same application software for word processing, even if they are all producing text for the same documentation project.

WORD PROCESSING FEATURES

Many technical writers do their own typing. Even physically handicapped writers manage to manipulate a keyboard with the aid of some device in order to create text. Indeed, some of the better known word processing applications provide special packs that aid in various cases of disability, such as help for those individuals with motion, hearing or low vision disabilities. One does not have to be a skilled keyboard operator, but it helps. Training in touch typing and plenty of practise will help cut down on errors, despite the availability of electronic spelling and grammar checkers (these are discussed later in this chapter). Word processors provide an important aid to writing. They help an author to reorganise blocks of texts and build the structure of a document easily, in almost any order, without the penalty of having to rekey any of the content. Since the text is stored on disk, it can be retrieved for editing as and when required and may be used as the basis of similar documents. The following paragraphs discuss a number of typical features of word processing software.

Editing Functions

Editing functions are those that relate to the creation, amendment and manipulation of the text that forms the content of the document. Text is input from the keyboard and is displayed on the screen as you type. A cursor indicates the position at which typed characters appear and this can be moved within the text area using special cursor movement keys on the keyboard or a mouse pointing device so that text can be inserted into or removed from any part of the document. There are usually two modes of typing; overtyping and inserting. In overtyping mode, any characters typed at the keyboard will replace any existing characters from the current cursor position, while inserting mode adds characters at the current cursor position, pushing any following text along in the document.

Text can be deleted using special function keys and blocks of characters, words, sentences and paragraphs can be selected and deleted, copied or moved within the document. A particularly useful editing feature is one that allows the author to search for words and phrases in a document and replace them with alternatives. The sophistication of such a feature depends on the word processor, but in some cases, searching for matching words or characters can be restricted to locating text of a particular style, type size, case, etc., using the whole of, or just part of a word or phrase. The search facility is useful because it can be used to locate technical terms, cross references and other important elements in a document. For example, if cross-references between different chapters cannot be specified until a document is completed, the author may enter a cross-reference during drafting of the text as 'see chapter x' or 'see also section x.x', etc. Once the structure of the text has been decided, a search may be done on 'chapter x' or 'section x.x.' so that the correct cross-reference numbers can be inserted. For a lengthy document, this can be far quicker and more thorough than trying to find all occurrences in printed draft by eye. (Having said that, more sophisticated word processors can even handle tasks such as automatic cross-referencing with ease.)

The replace function can be useful to make changes to words or phrases that are likely to change. For example, during development of a product, the product's name may be undecided. An author working on a draft of a manual to accompany the product may refer to it in the text as 'product x'. Once the name is agreed, a search can be made on 'product x' and automatically replaced with the correct product name. Other uses of search and replace might include altering the case of a particular word or term, for example, changing 'TERMINAL' to 'terminal'; changing spellings when editing transatlantic documents, for example, changing 'color' to 'colour'; replacing special format controls such as changing a single carriage return to a carriage return and line space; removing unwanted spaces by replacing occurrences of more than one character space with a single one; changing styles, for example, by searching for text in a particular typestyle and replacing the typestyle itself without altering the text; etc.

A summary of editing features include the following:

(a) input of new text from the keyboard;

(b) navigation of the document by being able to quickly move from one part of the document to another;

(c) deleting single characters, words, sentences, paragraphs, etc.;

(d) copying blocks of text for replication elsewhere in the document;

(e) moving a selected block of text to a new position in the document;

(f) searching for words, characters, terms, styles, commands, etc.;

(g) replacing text, styles or commands found by the search function with an alternative.

Spacing, Alignment and Other Formatting

Word processors provide the means to control the appearance of text automatically. Attributes that affect the format of a document, including spacing between lines of text, paragraphs, left and right margins, tabulation and indents, are generally handled automatically once the author has set up the required parameters. Tab positions can be applied to part or all of a document and can be changed or applied retrospectively so that a block of text can be realigned if necessary. Special tabulation features help to align columns of figures by using the position of the decimal point as the alignment key.

Other formatting features can be employed to control the way in which text is positioned in relation to other elements and where it appears on the page. For example, text used for a heading can be forced to begin a new page. If preceding text is removed or extended, the heading belonging to the next topic or chapter will therefore remain at the top of the following page, thus making editing easier to handle. Paragraphs may be 'anchored' to graphical objects or indeed, text may be formatted in such as way that a particular graphic, such as a line, shaded background or boxed border, forms part of the style of that particular text.

Because there is so much you can do with word processing packages as far as formatting is concerned, many offer a facility to set up 'styles'. These are a collection of formatting instructions that include the choice of typeface, size, alignment, spacing, tabulation and other characteristics that can be stored under a style name of your choice. Whenever you want to text in a document to appear in a particular style, the predefined style information can be applied to all or part of a document's text resulting in semi-automatic formatting.

This is particularly use in a technical publications environment where user documentation may have to adopt a variety of formatting styles according to the information being conveyed on the page at the time. The process of formatting text using styles can be done either during the writing stage, or as a separate operation. This is the kind of word processing that is similar in nature to desktop publishing; a topic which is covered in more detail in Chapter 6.

GRAMMAR AND STYLE CHECKERS

Facilities for checking the text of a document for spelling, grammar and style are commonplace in word processors and are also available as stand-alone applications which can check text generated with other software. Using such editing and proofing tools may help to improve the quality and readability of a document, though they are by no means foolproof and should not be considered as a replacement for actually reading through the document. There are a number of things that electronic proof-reading utilities can do, depending upon the application software being used. Broadly speaking, they fall into one or more of the following activities:

— finding and correcting spelling errors;

— correcting typographical errors;

— identification of possible grammatical or style errors;

— offering suggestions for improvement in style and choice of terminology;

— helping evaluate the readability of text;

— looking up synonyms, antonyms, and related words;

— providing statistical information about a document, such as a word count.

Spelling Checkers

These provide a useful initial check on spelling. They work on the basis of comparing each word in the document with an entry in a dictionary file. Unless specialist or edited dictionary files are used that include relevant technical terms, spelling checkers will stop at every such term, mnemonic or proper name. It is therefore important to ensure that an appropriate dictionary file is maintained. Some word processors allow more than one dictionary to be used during the checking process. Spelling checkers operate in a variety of ways, from scanning text files and tagging unknown words with a special character, to offering an interactive approach whereby the user can choose an alternative from a list presented or edit the misspelt word. For example, if a word is typed in as 'techbical', the spelling checker will probably offer 'technical' as a replacement. There may be several alternatives offered in some circumstances or none at all if the checker can find no match to any characters. Where a word is found that does not exist in the spelling checker's dictionary, the user can choose to ignore the advice, add the word to the dictionary or edit the text.

Interactive spelling checkers are far more efficient to use than those that simply mark-up unknown words. This is because in the latter case, the user must scan the text a second time to find the marks, then edit the text accordingly.

Although spelling checkers make excellent tools to help pick up silly typing errors and are particularly useful when preparing drafts, they are no replacement for a thorough proof-reading. Correctly spelt words in the wrong place will not be located. In the phrase 'there are two on three possibilities', a spelling checker will not pick up the fact the word 'on' should read 'or'.

Thesauruses

Some word processing software offer a built-in thesaurus to help find alternative words when the user gets stuck for ideas, or cannot think of the appropriate word to use in particular circumstances. They work as an on-line aide; that is, the user can invoke the thesaurus at any point while composing or editing text and, by highlighting or selecting a particular word, retrieve a list of alternatives. Some thesauruses are quite comprehensive and offer various possibilities within different contexts. Some may allow the user to tag or highlight their chosen synonym and have the program replace the original word automatically in much the same way as a spelling checker might replace a misspelt word with the correct one. The disadvantage of using an on-line thesaurus is that they tend to require a significant amount of computer disk space and memory in order to make an extensive dictionary readily available.

Grammar and Style Checkers

Grammar and style checkers are available as applications in their own right or built in to a word processor or desktop publishing system. They are designed to identify sentences that contain possible grammatical or style errors and suggests ways to improve them. Most proofing tools use one or more grammar and style rule groups. These are groups of checking attributes that can be used according to the type of text being checked. For example, the rules of style applicable to a technical document, such as a user manual, are likely to be quite different to those relevant to an informal letter.

Grammar checkers can only guess and advise because they are unable to recognise the context in which the document is written. They should also be used with caution and not taken too literally. In general, they are useful for identifying anomalies such as the improper use of plurals, capitalisation, missing verb subjects, and other basic grammatical deficiencies in the text. Some of the guidance given is subjective, especially in the use of split infinitives, which, in certain contexts, are not only permissible, but necessary to provide the right emphasis on a particular word or to clarify a sentence that might otherwise be misleading.

The following paragraphs illustrate the effect of a grammar checker on an extract from a user manual for a software application.

80 TECHNICAL DOCUMENTATION

```
The action enquiry form enables you to remind yourself or someone
else of an event/activity that is to be done on or by a particular
date and time. Such forms are only used to provide a brief description
of the action required and further information can be obtained by
clicking on the 'More' button which results in a pop-up window being
displayed that contains the full details of the action required. The
enquiry form can be accessed from either Action Enquiries menu, or
via the Action Link button attached to other sales contact management
information screens.
```

Figure 5.1 Example text, prior to employing grammar checker.

The text in the above figure was checked using the grammar checking software incorporated into Microsoft Word. The software checks each sentence in turn and stops to inform the user of each discrepancy or problem in turn. The following steps resulted from the checking process. In each case, the sentence being checked is presented and any problem words are highlighted in bold. A suggestion for improvement is provided and an explanation of the rule being employed is also available:

First sentence:

'The action enquiry form enables you to remind yourself or someone else of an **event/ activity** that is to be done on or by a particular date and time.'

Suggestion:

It is clearer to use 'and', 'or' or 'per' instead of a slash between words.

Explanation:

A slash is an appropriate separator for numbers such as fractions. Between words, consider using a word that expresses the exact relation you intend. Instead of 'Be sure to shake his/her hand', try 'Be sure to shake his or her hand'.

Second sentence:

'Such forms are only used to provide a brief description of the action required and further information can be obtained by clicking on the 'More' button which results in a pop-up window being displayed that contains the full details of the action required.'

Suggestion:

Consider revising. Very long sentences can be difficult to understand.

Explanation:

Long sentences often contain complex ideas that would be more clearly expressed in several shorter sentences. Within a document, sentences should vary in length for comprehensibility and rhetorical effort. Variety in sentence length and structure is an important part of good written style. Sentences with less than about 15 words are often direct and clear, but a document would be choppy and monotonous if all sentences were that short. Informative and technical texts may required sentences of 35 to 45 words for the expression of complex ideas, but paragraphs should not contain uninterrupted sequences of such complex sentences. Use shorter sentences for introductory, defining or concluding statements. Divide long sentences into shorter ones if the long sentence contains more than one major point.

Third sentence:

'The enquiry form **can be accessed** from either Action Enquiries menu, or via the Action Link button attached to other sales contact management information screens.'

Suggestion:

This main clause may contain a verb in the passive voice.

Explanation:

Excessive use of the passive voice can make a document unclear. Try rewriting the sentence using a more direct verb. 'Documentation will be included with the delivery' might become 'Each delivery will include documentation'.

As can be seen from the above examples, the information given about the content of the paragraph is quite extensive and the suggestions and explanations are informative. Clearly, using such a tool as this grammar checker on a document of considerable size would be very time consuming, but by using the facilities during writing, the author can adopt the suggestions that occur most often and gradually reduce the number of problems encountered. Notice, however, that the grammar checker failed to detect the absence of the word 'the' in the sentence 'The enquiry form can be accessed from either Action Enquiries menu, or via the Action Link button attached to other sales contact management information screens'. This should actually read 'The enquiry form can be accessed from either *the* Action Enquiries menu, or via the Action Link button attached to other sales contact management information screens'.

This highlights the shortcomings of any electronic proofing system. There are no spelling errors in the sentence but it is grammatically incorrect. Yet the grammar checker is unable to detect it. Only by reading through documentation thoroughly will it be possible to ensure that the content makes sense and conveys the correct information.

In Microsoft Word, as in other applications that include grammar checking facilities, additional information is also provided about the readability of the text. Figure 5.2 shows and example of the readability statistics that are provided as a result of checking the example paragraph in Figure 5.1.

The readability statistics information displays useful information such as the number of words, characters, paragraphs, and sentences; the average number of sentences per paragraph, words per sentence and characters per word; and other readability indexes for the document. In the readability section of the display, the software displays the percentage of sentences written in passive voice, as well as other readability indexes.

82 TECHNICAL DOCUMENTATION

Notice that it considers the example text to be 100% in the passive voice. Other indexes are based on special sets of rules and are explained below:

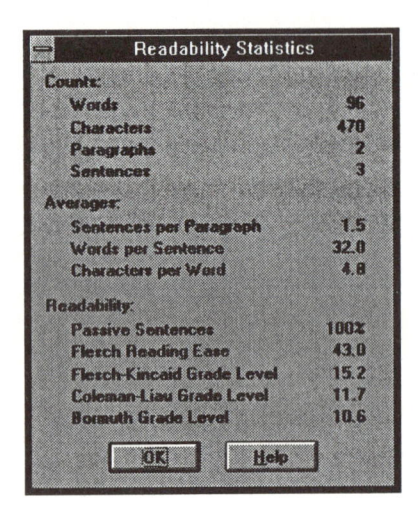

Figure 5.2. Display of the readability statistics of the example text.

The readability statistics information displays useful information such as the number of words, characters, paragraphs and sentences; the average number of sentences per paragraph, words per sentence and characters per word; and other readability indexes for the document. In the readability section of the display, the software displays the percentage of sentences written in passive voice, as well as other readability indexes. Notice that it considers the example text to be 100% in the passive voice. Other indexes are based on special sets of rules and are explained below.

Flesch Reading Ease

This computes readability based on the average number of syllables per word and the average number of words per sentence. Scores range from 0 (zero) to 100. Standard writing averages approximately 60–70. The higher the score, the greater the number of people who can readily understand the document.

Flesch–Kincaid Grade Level

Computes readability based on the average number of syllables per word and the average number of words per sentence. The score in this case indicates a grade-school level. For example, a score of 8.0 means that an eighth grader would understand the document. Standard writing approximately equates to the seventh to eighth-grade level.

Coleman–Liau Grade Level

Uses word length in characters and sentence length in words to determine a grade level.

WORD PROCESSING 83

Bormuth Grade Level

Uses word length in characters and sentence length in words to determine a grade level.

Clearly, some work needs to be done on the example text. Figure 5.3 shows the revised text and Figure 5.4 the improved statistical information. Notice how, by rearranging the words and sentences, the passive voice percentage has been reduced by 75%. Whether you think that the paragraph is any more informative is a matter of opinion. This must always be borne in mind when using grammar and style checkers. You may be able to improve on the statistics considerably by editing the text, but this does not necessarily result in information that is any easier for the reader to understand. Always consider the rule 'will the reader understand what I'm saying' and, in my opinion, you will have employed the most effective form of grammar and style checking you can.

```
Use the action enquiry form to remind you or someone else of an
event or activity to be done on or by a particular date and time.
The forms provide a brief description of the action. Click on the
'More' button to display a pop-up window containing full details
of the action required. Access the enquiry form from either the
Action Enquiries menu, or the Action Link button attached to other
sales contact management information screens.
```

Figure 5.3. Revised example text.

Readability Statistics	
Counts:	
Words	76
Characters	362
Paragraphs	2
Sentences	4
Averages:	
Sentences per Paragraph	2.0
Words per Sentence	19.0
Characters per Word	4.7
Readability:	
Passive Sentences	25%
Flesch Reading Ease	56.2
Flesch-Kincaid Grade Level	10.1
Coleman-Liau Grade Level	10.1
Bormuth Grade Level	9.7

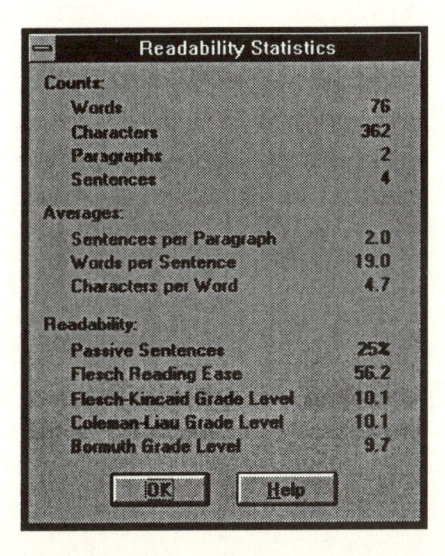

Figure 5.4. Readability statistics for revised example text.

AUTOMATIC REFERENCING AND INDEXING

With ever-increasing sophistication and function, word processors can control many of the laborious tasks associated with technical documentation automatically. Two of these tasks are cross-referencing and indexing. Cross-references tell readers where additional information is located in the same document or another document for a particular topic; for example, 'See Figure 6.1 on page 34.' Some word processors enable you to create a cross-reference to an item, by marking the item concerned first, then linking the item with its cross-reference. The main benefit of using such a facility is to aid future editing, because if the extend of the documentation changes as a result of amendments, such that a cross-referenced item appears in a different location, the word processing software can update all references for you.

Automatic indexing also offers considerable assistance to the author. Entries in the index are marked in the text, and any cross-referenced topics within the index can also be identifies to provide nested levels of indexing. For example, the subject 'printing reports' may appear both under 'printing' and 'reports' in the same index.

Indexes need to be designed to suit the user of the document. Consideration must be given to the way in which the user may wish to access the information contained in the document; the reader's level of subject knowledge is an important factor in deciding what to include. Indexes for product manuals may evolve from the function of the product and its application, with key words and terms being selected from the text that relate to particular features of the product. Indexes for maintenance, service and repair manuals need higher skills and experience to ensure that the organisation of the index, the selection of terms used and the presentational style are appropriate for the manner in which the manner will be used.

Indexes generated by a word processor must be checked for inappropriate entries; particularly where entries may be included automatically because the text is a particular heading level, key word or style. Where index entries can be selected manually from the text, the person responsible for selecting the entries must have sufficient understanding of the user's needs. It is not a practise that should be delegated to someone lacking that understanding.

The presentation of the index is almost as important as its contents. The entries should be neatly laid out, with related subjects clearly nested under the appropriate headings. Depending upon the page size and the average length of index entries, it may be possible to split the page into columns. Page references should be close to the corresponding entries, not right aligned with a large gap between them, otherwise they will be difficult to relate. Leader dots or lines should only be used in contents, not in indexes.

If the index is likely to be extensive, it is usual for the text to be in a smaller size than that of the document's main body text. This will also make it easier to fit the index into two or more columns on the page and therefore reduce the number of pages required. As helpful guide to index preparation for technical documentation can be found in the British Standard publication BS3700 (see Chapter 12, List of Standards).

SUMMARY

1. Word processors are universally employed through industry and commerce and have particular importance to technical authors, especially since most documentation is extensive and subject to constant amendment.

2. Most proprietary word processing applications offer similar features to define the layout and format of text as well as editing and manipulation controls. The distinction between word processing and desktop publishing is now hard to identify.

3. Some features of modern word processors offer technical authors considerable aid in otherwise laborious tasks. These range from proofing aids for spelling and grammar checking to sophistical tools for generating and maintaining cross-references and indexes.

6 Electronic Publishing

The introduction of computerised typesetting during the 1970s has changed the course of publishing in a significant way. Text can composed on a word processor or text editor, stored on disk and transferred to more complex editing and layout software for complete camera-ready artwork production or even direct to print.

DESKTOP PUBLISHING

Desktop publishing has brought the page make-up process onto the personal computer. By replacing the complex computers used for page make-up in the late-1970s with specially written software on modern business microcomputers, the entire process of typesetting, picture and graphics manipulation and camera-ready artwork production is available to everyone. With a wide range of high-quality output devices including laser printers, which can produce camera-ready artwork on plain paper or direct to film, the concept of in-house publishing is not only reality, but readily affordable to most organisations involved in the production of technical documentation.

For many newcomers to the subject, the purpose of desktop publishing is not always easy to distinguish from that of word processing. One tends to think of the creation, manipulation and output of text (otherwise called 'copy') from a computer system to a printer as word processing. The distinction between the two technologies is ever more difficult to define. Traditionally, the process of preparing text for publication using a computer, required the facilities for the keying in, editing, manipulation and arrangement of the words themselves. The copy preparation stage of any publication, whether it is a letter, report or brochure, is concerned mainly with the content, and not with the appearance or layout. A word processor can provide the means to do this, and by using the editing capabilities of a word processing software system, a draft document can be 'polished' ready for turning into artwork. Spelling checkers are a useful additional facility provided by word processors; and, for lengthy documents, the ability to be able to handle large amounts of text — which can be typed in quickly and efficiently — means that a computerised word processor makes an ideal tool for the job.

A desktop publishing system is largely concerned with the appearance of the text, its positioning on the page, the size and style of the characters, the inclusion of pictures and other illustrations; in fact with all the elements of what are collectively described as page make-up. A desktop publishing system is basically a page make-up utility. Since most page make-up systems display their layout on the screen in much the same way as the document will ultimately appear when printed on paper, the computer needs to do a lot of processing to display text in different styles and sizes — and with the display of graphics, such as photographic images, even more. Simple word processors do not concern themselves with the appearance of the text as such, even though they may be capable of printing the output in various styles. Instead, the screen display tends to be much simpler again, concentrating on the content rather than appearance.

The consequence of this difference between the two is that desktop publishing systems do not make very efficient systems for original text entry. They are slow to respond to the keying in of new text because they take time to present the styles and sizes of the page layout on the screen while the text is entered. For original text preparation then, standard word processors are preferred. Most desktop publishing systems rely upon the copy having already been prepared on a word processor. The text already stored in magnetic form on a disk can then be 'read' by the desktop publishing software, and styles can be attached to the text to define the page layout.

The same process applies to the manipulation of graphics. Desktop publishing systems, while they are able to cope with presenting and printing graphics such as line drawings, scanned photographs and the like, are not in themselves very suitable for the origination of the graphic images in the first place. Like text, the desktop publishing system relies upon a separate facility for the production of the graphic images — a graphics software system, for example, or a special piece of equipment used to convert a photograph into a computer image using electronic means.

So a desktop publishing system is distinguished from a word processor on the basis that it acts as a tool to place the elements of a page layout (text, graphics, logos, etc.) within a predefined print area, and to define the styles and sizes of the elements in order to maximise the quality of presentation and to most effectively convey the message contained within the text.

There is a slight confusion, however, over the fact that the result obtained from using a desktop publishing system, if specifically designed for page make-up, can in some cases be equally well produced using a sophisticated word processor. Such word processors are those that have taken the facilities of the basic system of text entry, editing and manipulation, and added extended features to control the style and presentation of the pages beyond that of their more traditional counterparts. The subject of different types of desktop publishing systems is discussed in more detail in later in this chapter

Page Make-up and DTP

The illustration in Figure 6.1 emphasises the basic page make-up principle of desktop publishing — the manipulation of various design elements and text to make up a whole page layout. Before technology permitted this work to be done using a computer, these separate elements had to be manually positioned on a page. Typeset text cut into appropriate lengths, pictures and other graphic elements, such as rules and boxes, all had to be pasted onto a layout board (a piece of cardboard, often with a fine graph-like grid for accurate positioning which did not reproduce when the artwork was photographed for platemaking).

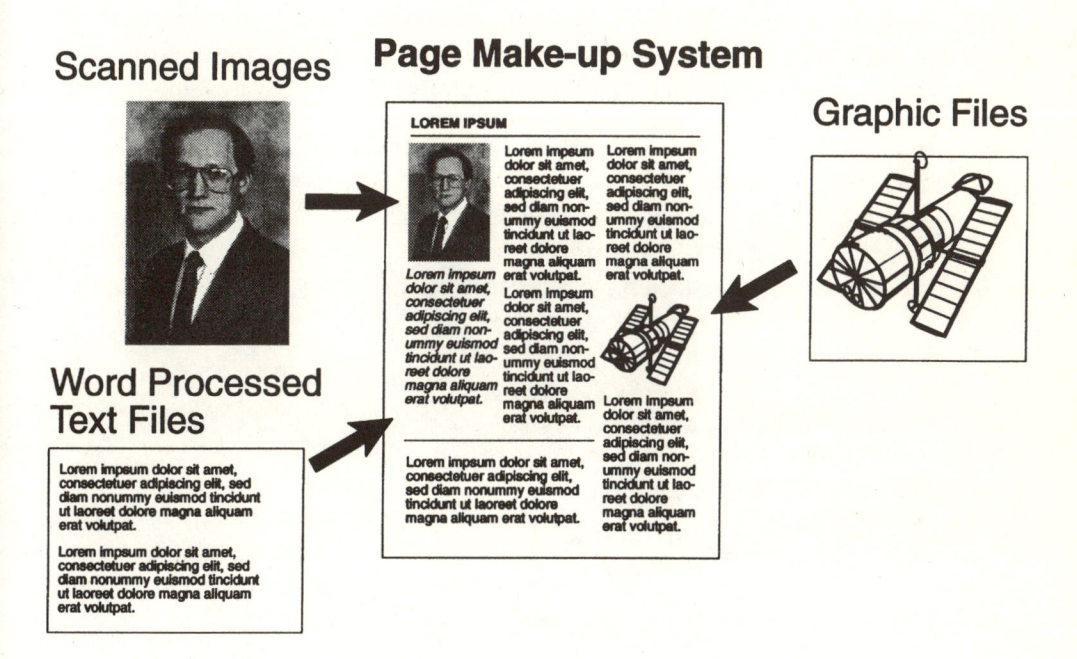

Figure 6.1 Principles of page make-up.

For those print and design establishments that do not yet use the page make-up facilities of a computer-based system, this method is still adopted. It is also common to find a combination of these techniques, where the printed output of a desktop publishing system provides the basic layout and organisation of all the text elements; and where graphics such as photographs and logos are pasted on afterwards.

Style Control

Desktop publishing systems all provide control over the style of text and page layout. Some provide finer control than others. Depending on what you are trying to achieve,

90 TECHNICAL DOCUMENTATION

the facilities may or may not be adequate for the job. The decision-making process to determine which desktop publishing page make-up system to use requires careful consideration of your intentions.

All DTP systems will provide common basic functions. These include the ability to select a style and size of typeface. The typeface is the design of the characters which affects the appearance of the text. Typefaces are designed to be attractive but they do require careful selection so that the choice does not impair the readability of the document. The typeface then, is the style of the characters used for the text. Typefaces have their own names and design characteristics, and a complete set of characters in a particular typeface — for example all letters, numbers and special characters — is called a font. The illustration in Figure 6.2 shows a selection of typefaces. Desktop publishing systems offer the operator a choice of typefaces and sizes to use in a document layout, though the variety offered will depend upon many factors, including the type of printer and any other output device being used, the capabilities of the software, etc.

Arial
Arial Bold
Arial Italic
Arial Bold Italic
Avant Garde
Avant Garde Bold
Avant Garde Italic
Avant Garde Bold Italic
BakerSignet
Baskerville
Bauhaus
Bembo
Benguiat
Blippo
Bookman
Broadway
Candida
Century Old Style
𝕮𝖑𝖔𝖎𝖘𝖙𝖊𝖗 𝕭𝖑𝖆𝖈𝖐
Courier
Courier Bold
Courier Italic
Courier Bold Italic
Cushing

Figure 6.2 Selection of typefaces.

One document may contain more than one typeface; for example, headings will often be of a larger type size and sometimes in a different style to those of the body text of the document. Using different typefaces adds character and interest to the appearance of the document. Even something as basic as a mailshot letter can be made more acceptable and interesting by the appropriate selection of typefaces and sizes. Care has to be exercised, however, so as not to over-use such facilities. But with appropriate training and experience, an operator can improve upon the appearance of documents that were previously confined to the typewriter or standard word processor.

Apart from the style of the type used in a document, a page make-up system will provide facilities to set up tabs, margins and columns, into which the text can be fitted to make the organisation of the various text elements interesting and informative. Newsletter-style documents will probably tend to copy the style of a newspaper, with text placed in two or more columns on a single page, unlike typical book style which has only one single column. Multiple-column manipulation of text is a unique feature of desktop publishing page make-up systems, one which could not be adequately provided for by word processors.

There is a fundamental difference in the use of spacing with typesetting from any typewriter. With typesetting, all the characters have an amount of horizontal space which is proportional to the actual width of the character, so over a large amount of text, it is quite noticeable how much less space is needed to fit text into a predefined space. Typewriters and word processors that, on the other hand, use daisywheel printers tend to use non-proportional spacing. So the letter 'i', for example, will occupy the same amount of horizontal space as the letter 'm', even though the 'm' is a wider character. All desktop publishing systems offer proportional spacing of text and this is one of the fundamental differences between typeset text and non-typeset text.

```
This paragraph of text has been set using a non-propor-
tional typeface. Every character has the same amount of
horizontal space. This makes it easy to line up columns of
figures, for example, when producing a table, but the typeface
does not make optimum use of the space on the page.
```

Graphics

In addition to the control over the appearance of the text, desktop publishing systems generally provide a number of features that enable graphic elements to be included in the text. These include rules (vertical and horizontal lines) boxes, shading, etc. These elements can be used to improve the appearance of the page further; for example, a rule might be used between the columns of text in a report to separate them. Some desktop publishing systems provide a number of functions for limited drawing capabilities, but more complex pictures generally need to be produced using a different system, and then 'imported' into the page make-up system to be positioned on the page as required.

Output Devices

The output devices supported by various desktop publishing systems can vary. Most now offer output to a laser printer — a high-quality printing device that allows typesetting of text at a fraction of the cost of phototypesetting, though the phototypesetter remains the ultimate quality. This book has been typeset on a 600dpi laser printer. Inkjet and thermal printers also feature as output devices, and all are capable of handling different typefaces and graphics.

Methods Employed

There are several ways in which page make-up can take place. In some cases, a certain amount of the style control of a document can be predefined in the original word processed text before it reaches the desktop publishing software. In other cases, a desktop publishing system will take raw text and every attribute to do with the style of the text and the page layout has to be added afterwards. How various desktop publishing systems actually carry out the process of page make-up depends upon the 'type' of system.

CATEGORIES OF DTP SYSTEMS

Desktop publishing can be broken down into a number of categories. These relate to the degree of capability of the system and therefore ultimately the purpose for which they are intended. As a general guideline, the different categories are separated into four groups:

Group I — Word processor-based systems

Word processors that can produce a varied layout, can control the size and style of the typeface, and with sufficient intelligence to produce output on almost all output devices.

Group II — PC-based page make-up systems

Simple and advanced page make-up software systems, designed to run on standard personal computers, which enable text and graphic elements from external systems (such as word processors and drawing software or scanners) to be combined in a page layout, and to provide control over the appearance of output which may be on any device.

Group III — PC-based electronic publishing

Also page make-up software systems designed to run on standard personal computers, but offering a much higher level of function than those in Group II, generally by providing better typographical control of the design and the manipulation of page elements. Systems in this group are capable of output to any device but may include sophisticated composition capabilities, colour separation of text and graphics and the finest degree of typographical control.

Group IV — Dedicated electronic publishing systems

Systems offering a similar level of function to those in Group III, but on hardware which is dedicated to the purpose of desktop publishing. These systems tend to have an advantage for professional users, by providing specialised keyboard layouts to facilitate manipulation and selection of specific typographical functions, the use of paper-sized screens (A4 and A3) and tailored software features.

The systems in each of these groups provide some overlapping features. It is not always easy to differentiate between adjacent categories of desktop publishing, though the differences between, for example, word processing-based DTP and dedicated electronic publishing systems are usually considerable.

WORD PROCESSOR-BASED DTP

Word processors capable of producing a certain degree of layout control over standard typewriter-style document production, offer features very similar to those of basic desktop publishing systems. Consequently, in this group of software systems, it is difficult to differentiate DTP from word processing. Such word processors are able to take advantage of the capabilities of output devices such as laser printers and phototypesetters, to produce character styles of different typefaces and sophisticated graphics, possibly in colour. Some offer the ability to produce documents in almost any format and size, allow picture images to be included and other similar page make-up functions, but they retain the facilities that make word processors most suitable for original text entry and manipulation — such as spelling checking, search and replace of words or phrases, cut and paste (moving blocks of texts within the document) and other similar editing features — plus office productivity features like mailmerge. In order to control the style of the printed output, such systems may use a style sheet system in which the control of font, typestyle, size and position of text and other elements can be applied to text by simply attaching a style name.

Such systems are not necessarily cheaper than desktop publishing software, but since most page make-up systems rely on the use of some text editing or word processor features anyway, the overall cost may be less. For staff used to the operation of word processors, such systems are easier to operate in the short term, since they use familiar methods of text manipulation. The setting of tabs and margins, page lengths, headers and footers, for example, is not unlike that with any standard word processor. If the application for which the system is to be used is limited to the production of high-quality letters and reports, then the word processor based systems can be most adequate.

With operating platforms, such as Microsoft Windows and Apple Macintosh, word processors design for these environments are able to show accurate representations of the page layout and styles on the screen (as if the screen were the printed page). While some word processor-based systems offer this facility in the form of a 'preview' option (whereby the text is edited in an unformatted mode and then displayed like the printed page as an option), page make-up systems tend to present the document on the screen in this manner as a matter of course.

PC-BASED PAGE MAKE-UP

As previously explained, the basic principles of desktop publishing revolve around the process of page make-up. In this context, the basic elements that comprise a document, the text and any associated graphics, are produced externally to the page make-up software. Text will have been generated on a word processor, and stored as a file on the computer disk. The file is then read by the page make-up software and the text placed into a page layout accordingly. Different word processors store their text in different formats, i.e. any information (indents, page breaks, etc.) stored with the words themselves may be encoded in the file, and the system of encoding used by one word processor will not necessarily be the same as that of another.

Consequently, page make-up systems need to be able to read or recognise a variety of text file formats. Those of the most commonly sold word processors are generally catered for. Since many word processors can output a 'straight text' file known as an ASCII file (one which has no special format codes at all), and page make-up systems can always read this, there is usually little difficulty in transferring word processed text to the desktop publishing system.

In a similar way, pictures produced by graphics software packages store images as files on the disk. Like text files, different graphics systems store picture images in different formats. Again, a page make-up system must be able to recognise and read the various formats in order to be able to include an image within the page layout. Once the elements of the page layout are within the control of the desktop publishing software, they may be manipulated to produce the final page layout. For text, this means specifying typefaces, styles, sizes and spacing for headings, paragraphs and the like. For graphics this may mean sizing and positioning.

The underlying structure of the page layout may be defined separately without the need of the text and pictures that will be included with the final printed document. This process is called defining a template. A template comprises all the attributes that describe how the page elements will be arranged and displayed: for example, the size of columns into which text will be placed, the size of the page itself, the typeface to be used for headings, subheadings, paragraphs, etc. Once a template is defined, the text and pictures can be called into the page make-up system and after final adjustments and manipulation, the complete document can be printed out. Because templates hold information relating to the design and layout of the document, but do not necessarily include any 'content', they may be used for the production of many documents that require a similar style.

Page make-up systems store the complete document as a 'publication'. How this is actually done varies from one product to another. For example, the original text (word processed) and graphics files may be left intact, and the completed document stored as a separate file. Alternatively, the publication may be stored in a number of separate files — text in one, graphics in another, template information in a third, and so on.

Some of the most popular page make-up systems provide a facility to enable a word processed document to be prepared for page make-up, to an extent which reduces the amount of work that would otherwise need to be done on the document to assign the various style attributes. For example, the following attributes may be assigned to a main section heading of a report:

Text:	centred
Typeface:	Helvetica
Style:	bold
Size:	18 point
Spacing beneath:	2 lines
Special:	begins a new page

These, and other, attributes can be assigned to a simple code or name; for example, H1 for a first level of heading.

Now, within the same document, a subheading may have slightly different attributes:

Indent from left:	10 mm
Typeface:	Times
Style:	bold
Size:	12 point
Spacing beneath:	1 line

As with the main heading, this heading may be associated with the code H2 — a second level heading. The body of the document — the paragraph text (often referred to in page make-up systems as 'body text') — will have yet another set of style attributes:

Indent from left:	10 mm
Typeface:	Times
Style:	Roman
Size:	10 point
Spacing beneath:	0

This may have a name such as 'body text'.

Having assigned the styles to the template and given them names or codes, the text within the word processed document can be prepared to include the style information. For example:

[H1]Section 1 - Sales Report

[H2]1st Quarter Performance

[body text]Compared to the results of previous years, the sales in the same quarter this year have been encouraging. With three additional representatives in the field, we have been able to offer more personal support services to existing customers as well as extend our market penetration in those areas targeted in the sales budget report...

The codes and names used, such as H1, H2, body text, etc., are generally referred to as 'tags'. They are enclosed in non-standard brackets or are preceded by a little used character, such as @, so that the desktop publishing system can differentiate them from the normal text. The names used will match the names of the styles defined in the page make-up software, so when the code [H1] is encountered, for example, it knows that the following text is to adopt the styles set up under that name. The style will then only change when another style code is encountered.

Figure 6.3 shows the example above after processing and output from the page make-up system. There are various terms associated with this technique — generic coding, text style code embedding, document tagging, for example. The principle is particularly helpful for the copy preparation of long documents which have a fairly simple style. If the document is complex in layout and design, using such preparation techniques is not really practical.

Section 1 - Sales Report

1st Quarter Performance

Compared to the results of previous years, the sales in the same quarter this year have been encouraging. With three additional representatives in the field, we have been able to offer more personal support services to existing customers as well as extend our market penetration in those areas targeted in the sales budget report...

Figure 6.3. Generically coded text example after processing.

ELECTRONIC PUBLISHING

Electronic publishing is a term often used as a substitute for desktop publishing. It does, after all, describe the same process of publishing with a computer. But the term is also used to express the difference between 'professional' publishing and general use. The general application of desktop publishing tends to cover the use of PC-based page make-up software, used on generally available personal computers which may also be used for other applications, and where publishing is not the primary concern of the organisation or department using the system. Desktop publishing is, in this respect, an embellishment to word processing, and occasionally employed to produce good-looking reports, overhead transparencies, price lists, letters and other sundry business-like documents.

Similar systems would also be used by educational establishments for teaching the basics of the technology and for general-purpose publication use, again, not a primary function of the organisation. By contrast, the professional sector of the market is that in which desktop publishing performs an essential function for the organisation or department. For example, if a company is in the business of printing, design or publishing, then its primary function will revolve around (or significantly rely upon) the production of artwork in one form or another. A department in a large organisation, whose responsibility is the publications and printing for the organisation as a whole, would likewise find desktop publishing playing a major role.

The significance of the resource of desktop publishing to an organisation will have an effect on the choice of system employed. As a general rule, electronic publishing systems are sophisticated desktop publishing systems aimed at users who are familiar with printing and publishing.

From the layperson's point of view, the difference between PC-based page make-up systems and electronic publishing may not be obvious at first, or indeed, may never be obvious. So what characteristics help identify electronic publishing from the wider, more general market of applications?

Here are a few pointers:

— they may be as much as twice the price of the most popular middle range packages;

— they support much more sophisticated output devices which will include one or more industry standard phototypesetters, or output direct to film or plate;

— they allow much finer control over the page layout; including spacing, positioning and other typographical aspects not essential for general-purpose use but which may be critical for professional typesetting;

— they offer full colour separation and graphic image control;

— they are often sold as a cheaper and more adaptable alternative to conventional phototypesetting 'front ends'.

DEDICATED SYSTEMS

These are systems that bundle the electronic publishing software of the type described above, with special hardware designed specifically for page make-up. The hardware, for example, might incorporate an A4 or larger screen and special keyboard with keys for unusual characters and typesetting commands.

These systems are really just alternatives to the conventional phototypesetting systems, but may employ a laser printer as either the main output device, or for proofing purposes. The advantages of such a system include:

— specialised hardware and software facilities built-in;

— complete system from one supplier, making support and maintenance easier to manage.

Disadvantages include:

— less adaptable to a changing and developing market since all technology is tied to one supplier;

— costly for additional terminals compared to PCs;

— often occupies more space than PC-based systems.

DATABASE PUBLISHING

Database publishing is really an extension of desktop publishing, except that, instead of using word processed documents, graphic images and other page elements for the source of the content of the document, they provide a means of formatting information stored in a database for printing in a particular style or layout.

Database systems can include any software-based application that stores and retrieves data. This may be an information system containing anything from a company's accounts to government statistics. Information within databases is stored in a hierarchical structure of files, records (or tables) and fields. Each field is one element of data, such as a person's name. Each record or table represents a group of related fields, such a company record that includes fields for company name, street address, town, county, telephone number, contact name, etc. Files contain the records that comprise the database, so an information system that stores, for example, all UK companies and organisations by industry sector, may have different files for different industry sectors. Thus, all service-based company records may be stored in one file, all financial institutions in another, and so on.

By employing the use of a database publishing system, it becomes possible to print reference and directory documents on the basis of templates or style sheets that can attribute particular characteristics to individual elements or fields within the database. When the relevant information is extracted in the order required, the database publishing system can compile a directory of companies, for example, with each company name in one type style and position, address in another, contact and telephone details in yet another, and so on. This process of automating the publication of database information is relevant only to specialist publication departments and is in itself a significant topic beyond the scope of this book.

SUMMARY

1. Desktop publishing (DTP) technology is readily available on personal computers and this has provided technical authors with the means to carry out the page design and make-up process.

2. DTP systems are available in different levels from simple software applications for personal computers through to dedicated systems with specialist hardware.

3. Many modern word processing packages offer most of the features of a true desktop publishing system, though more typographical control is usually afforded by the more sophisticated DTP systems.

4. DTP allows the integration of word processed text files, graphics, scanned images video capture and other elements within the same document.

5. The control of text styles is often provided through stylesheets or templates. These determine the characteristics of the text such as the choice of typeface, style and size as well as other typographic attributes such as level of indent, leading, weight, colour, etc.

6. Text generated on word processing packages can be tagged with stylesheet or template names so that, when imported into a DTP system, the appropriate characteristics are applied automatically to each level of heading and text.

7. Database publishing provides the means to control the typesetting of information extracted from a database. This technology is commonly employed when creating artwork for reference documentation, parts lists, directories, etc.

FURTHER READING

Jones, G. *The Desktop Publishing Companion* (Sigma Press, 1988).

Lang, K. *The Writer's Guide to Desktop Publishing* (Academic Press, 1987).

Marlow, A.J. *Desktop Publishing, A Management Report* (NCC Blackwell, 1990).

Marlow, A.J. *What is Desktop Publishing?* (NCC Blackwell, 1990).

Worlock, P. *The Desktop Publishing Book* (Heinemann, 1988).

7 Graphics and Illustration

All courses and instruction books on technical writing will emphasise the benefits of illustrations in your documentation. 'A picture is worth a thousand words' as Confucius said. The possibilities for illustrating documentation have become significantly widened in the advent of desktop publishing technology. With high-quality output devices like laser printers and plotters, and a wide range of graphics manipulation and drawing software for personal computers, the need for hand-drawn technical illustrations is dwindling rapidly and the illustrator must be adept at using this new technology in order to remain competitive with the increasing production of illustrations in-house.

A good line diagram can be useful in amplifying and clarifying main points in the text. Apart from its obvious benefit in this respect, the use of illustration material in a technical document helps break up the text which might otherwise seem monotonous in a large document, and thereby retain the reader's interest.

Just how much illustration work to use will depend on a number of factors, not least of which will be what facilities and budget are available for a particular documentation project. Some illustration work, especially in documentation for mechanical or engineering projects, will require specialist illustration and technical drawing skills, while software documentation can often be illustrated by any reasonably competent computer user given the considerable help offered by graphics application software.

Organisations that require design illustrations may need to employ the use of specialist CAD/CAM systems. These are a combination of computer hardware and software that provide the right kind of 'tools' to build detailed and minutely accurate plans and engineering drawings. Some offer the ability to generate 3D projections of prototype models and the more sophisticated systems can also rotate, skew, colour, shade and animated an object from one original design.

For more modest illustration requirements, once again, the personal computer provides solutions that enable authors to create effective and imaginative illustration material and integrate these through a word processor or desktop publishing system with the text of a manual or report. It is worth looking briefly at some of the main technological facilities that are available to technical authors to assist them in illustration work as well as considering some of the implications for reproduction of the material.

GRAPHICS AND DRAWING SOFTWARE

Software for graphics has gone through quite a boom in the past few years. This has largely been due to the impact made with the introduction of desktop publishing and more versatile word processing systems. Software that handles the production of graphics ranges from simple drawing packages that allow free-form drawing of lines, boxes, circles, etc. with text and many incorporate colour. Other systems offer far more control over graphics objects. These not only allow any free-form drawing, but also enable the user to manipulate objects; straightening wavy lines, skewing and rotating boxes, shading, resizing, inverting, overlaying, etc.

Many desktop publishing systems and word processors offer a limited amount of drawing facilities within them. To draw a box around some text does not normally require the use of any other software application. Simple charts and flow diagrams can often be generated with the system and included in with the appropriate text. More complex graphics produced with external graphics software can be imported in a variety of formats to be included in the appropriate part of the document. Graphics can sometimes be anchored to text so that when a document is edited, the illustration remains in place with its caption or associated paragraph. Alternatively, a graphic can be fixed in position on the page so that text flows around it or jumps over it.

Where simple diagrams or illustrations are required, the advent of 'clip art' has provided authors with ready-made solutions. Clip art is copyright-free icons, images, symbols, backgrounds and other graphical objects, either in colour or monochrome, that can be imported into a word processing or desktop publishing document. Some clip art can be quite complex and even stunning while others are simple and crudely drawn. The choice available is considerable and some drawing application software is supplied with massive libraries of clip art on disk or CD-ROM. Other clip art and library photographs can be obtained through international computer networks such as Internet and CompuServe. Figure 7.1 shows examples of clip art images.

Line copy is any image that is made up of solid black, with no gradation of tone. Some of the examples of clip art shown in Figure 7.1 are line art images such as the arrow, the aeroplane and the gears. Half-tones are images with continuous tones of grey. The clip part image of the pipes in Figure 7.1 is a half-tone image. Black and white photographs are also half-tone images. Colour images also have tones, which are converted to greyscale if the output device cannot reproduce the original colours. Some of the clip art images in Figure 7.1 are colour, such as the computer keyboard, diskette,

GRAPHICS AND ILLUSTRATION 103

Figure 7.1. Examples of clip art images.

no smoking sign but the colours are reproduced in this book as greyscales. Many graphics software packages provide facilities to control the half-tones, colours, brightness and contrast of graphic images so that an optimum combination of elements can be achieved according to the method of output.

Half-tones and colour images may be more attractive that line art images but there is a disadvantage. The space occupied by graphic images in terms of computer memory and disk space can be considerable. A complex colour graphic or detailed half-tone image may stretch the capabilities of even the most powerful personal computer and output device. Line art objects can be simple or complex but will require less powerful computers for their manipulation and do not occupy as much disk space.

Image File Formats

Most word processors and desktop publishing applications can handle the import of graphic files in a variety of formats. Different formats are generated by different applications and some are more suitable for particular types of images than others. For example, not all file formats can support full colour graphics and others are do not retain sufficient information for detailed image processing that may be required for half-tones. Some of the most common graphic file formats are listed below.

Note that different image file formats hold different quantities of colours. The number of bits per pixel that the format is capable of supporting indicates the number of colours supported:

— 1 bit per pixel refers to a image with up to 2 colours;

— 4 bits per pixel refers to a image with up to 16 colours;

— 8 bits per pixel refers to a image with up to 256 colours;

— 16 bits per pixel refers to a image with up to 32,768 colours;

— 24 bits per pixel refers to a image with up to 16,777,216 colours.

BMP/DIB/RLE File Formats

These files are known as 'Device Independent Bitmap' files, or 'DIB's'. They exist in two different formats:

OS/2 Format Images — These are saved using this format may be used with OS/2's Presentation Manager.

Windows Format Images — These are enhanced DIB file formats released with Microsoft Windows.

Although their file name extensions are different (.BMP, .DIB or .RLE), the files themselves are the same whether in used OS/2 or Windows. BMP files can be created by most Windows-based graphic drawing or painting software applications. DIB files can be used as image files in the Windows environment and can also be applied to computer multimedia systems. RLE files are Windows 'DIB' files that are compressed to save disk space using an RLE compression routine.

Format	bits per pixel
BMP-OS/2-RGB	1, 4, 8, 24
BMP-Windows-RGB	1, 4, 8, 24
BMP-Windows-RLE	4, 8
DIB-OS/2-RGB	1, 4, 8, 24
DIB-Windows-RGB	1, 4, 8, 24
DIB-Windows-RLE	4, 8
RLE	4, 8

GIF File Formats

GIF files create the smallest possible image files and are therefore popular for uploading and downloading from electronic Bulletin Board Systems (BBS) like Internet or CompuServe where the size of the image file is important (the larger the file, the longer it takes to transmit and the more line costs are incurred).

GIF files may use an encoding method referred to as 'interlacing'. When an image is saved by using four passes instead of just one, it is called interlacing. On each pass, certain lines of the image are saved to the file. If the program decoding a GIF file displays the image as it is decoded, the user will be able to see the four passes of the decoding cycle. This will allow the user to get a good idea of what the image will look like before even half of the image is decoded. All the GIF formats support 1, 4, 8 bits-per-pixel and may contain more than one image.

PCX File Formats

These were originally created for use with the Zsoft Paintbrush program and is supported by more applications than any other format.

Format	bits per pixel
PCX Version 0	1
PCX Version 2	1, 4
PCX Version 3	1, 4
PCX Version 5	1, 4, 8, 24

TIFF File Formats

The Tagged Image File Format (TIFF) is an industry standard file format designed to provide considerable flexibility and offer various possibilities of how a TIFF image is saved. The TIFF format differentiates between types of images: black and white, greyscale and colour. If you need to generate graphic files for half-tones (including photographs) or complex colour images, the TIFF format is the most suitable and is accepted by almost all applications including desktop publishing and word processing software. The TIFF format can use one of six encoding routines including No-compression, Huffman, Pack Bits, LZW, Fax Group 3, and Fax Group 4.

Format	bits per pixel
TIFF-No Compression	1, 4, 8, 24
TIFF-Huffman	1
TIFF-Pack Bits	1
TIFF-LZW	4, 8, 24
TIFF-Fax Group 3	1
TIFF-Fax Group 4	1

IMG File Formats

These were designed to work with the GEM environment in which, among others, the application Ventura Publisher worked before the advent of the Windows version. Some desktop publishing applications support the importing and exporting this, now dated, graphic file format. The IMG format supports 1, 4, 8 bits per pixel.

MAC File Formats

MAC files come from the Macintosh program MacPaint and there are libraries of clip art available in this format. They can be used in a PC environment. MAC format files support 1 bits-per-pixel.

TGA File Formats

The Targa TGA format was developed by Truevision for their Targa and Vista products. It is an industry standard although not as widely supported as PCX or TIFF formats. TGA files may be saved as non-compressed or compressed (run-length encoded). TGA files support 8, 16, 24, 32 bits per pixel.

Vector File Formats

Apart from the formats described above, there are a number of other formats that are generated by specific application software that are categorised as 'Vector' or 'Object-based' graphics. These include CGM (computer generated meta) and CDR (Corel Draw) file formats in which shapes are represented by lines and curves as opposed to pixel based bitmap images like PCX files which are generated by paint software and scanners.

INCLUDING GRAPHICS IN DOCUMENTS

There are various ways in which graphics can be included in technical documents. If a desktop publishing or word processing system is being used, images can be generated from a number of sources:

— drawn by hand, as in the case of plans or engineering drawings;

— created using a CAD/CAM or graphics application software package;

— captured from still or motion camera;

— scanned from flat artwork or slide.

Graphics drawn by hand can either be 'pasted' onto a space provided on the artwork for the page or scanned or photographed to create a graphic file that can be imported into a desktop publishing or word processing package. Graphics generated by CAD/CAM or drawing software can be transferred to the pages in the desktop publishing system in one of two ways. Some systems, such as the Apple Macintosh and Microsoft Windows operating environments, provide what is called a 'clipboard' facility, which stores text and graphics temporarily for transfer to another application or for moving or duplicating elsewhere in the same document. The image is copied onto the clipboard, which is, in fact, a portion of computer memory allocated for storing. Once stored in the clipboard, the desktop publishing application can be loaded and access can be made to the clipboard to 'paste' its contents into a designated position on the page. A more common way of transferring the image is to store the drawing on the disk as a file, perhaps using one of the file formats described earlier.

Once an image has been placed into the document, it effectively becomes a graphics object in its own right, which can be moved around the document for repositioning, tilting, rotating, sizing and cropping. Sizing of graphic objects is simply a facility which enables the image to be scaled to a new size or dimension. This is particularly useful if the image is to fit a certain space on the page and the original drawing bears no direct relation to the space available in the publication. Most desktop publishing systems permit graphics to be scaled, though, depending upon the quality and resolution of the original image, the picture can lose some of its clarity if scaled too large or reduced too small from its original size. Cropping is another way of fitting a picture into a specific area, but instead of reducing the entire image, parts of it may be cut away around the edges in order to fit the space available.

Photographs may be included in desktop publishing documents so that they can be printed with the text, rather than pasted onto the paper afterwards. This is particularly beneficial if the printer is to reproduce many copies of a document for distribution. In order to incorporate a photograph into a document, the image may either be scanned or photographed with a digital camera in the first place. Another method is to capture a single frame (or still) from a video. Multimedia computers provide a wide range of video capture facilities, including direct capture from TV, and it is possible to obtain libraries of photographic images on CD-ROM. It is important to recognise the copyright ownership of such material.

Desktop and hand-held scanners can play a useful part in an in-house publications department as both line art and photograph images can be converted into image files for editing and importing into a document. Scanners work by simply converting the continuous tones of a black and white photograph or the colours of a colour image and converts them into one made up of a series of coloured or black dots, to approximate the varying shades and tones of the original. Once translated into a pattern of dots, the image can be processed through a computer and stored on the disk as a graphics file in one of a variety of formats according to the degree of detail and colour reproduction required.

108 TECHNICAL DOCUMENTATION

Some graphics application software systems provide special effects to alter the appearance of a scanned or imported image. Images can be merged, distorted, morphed (where one image gradually changes into another) and retouched to produce a result that may not even bear any resemblance to the original.

POSITIONING OF ILLUSTRATIONS

Whatever the source of a graphic image, once it is captured within the desktop publishing system, it can usually be easily manipulated to make positioning and sizing suitable for the publication in which it is intended to be printed. Already mentioned are the scaling and cropping features. Apart from these, many systems allow text to be flowed around a graphic image. This is useful when the image is smaller than the width of the text column, are can simply be used to create an interesting visual effect. The system may provide control to determine the shape around which the text will flow, which may be different from the shape of the original image. See Figure 7.2 for an example.

Lorem ipsum dolor sit amet, consectetuer adipiscing elit, sed diam nonummy nibh euismod tincidunt ut laoreet dolore magna aliquam erat volutpat. Ut wisi enim ad minim veniam, quis nostrud exerci tation ullamcorper suscipit lobortis nisl ut aliquip ex ea commodo consequat. Duis hendrerit in vulputate velit dolore eu feugiat nulla et iusto odio dignissim qui delenit augue duis dolore te dolor sit amet, consectetuer nibh euismod tincidunt ut volutpat. Ut wisi enim ad tation ullamcorper suscipit commodo consequat. Duis hendrerit in vulputate velit dolore eu feugiat nulla et iusto odio dignissim qui autem vel eum iriure dolor in esse molestie consequat, vel illum facilisis at vero eros et accumsan blandit praesent luptatum zzril feugait nulla facilisi. Lorem ipsum adipiscing elit, sed diam nonummy laoreet dolore magna aliquam erat minim veniam, quis nostrud exerci lobortis nisl ut aliquip ex ea autem vel eum iriure dolor in esse molestie consequat, vel illum facilisis at vero eros et accumsan blandit praesent luptatum zzril delenit augue duis dolore te feugait nulla facilisi. Nam liber tempor cum soluta nobis eleifend option congue nihil imperdiet doming id quod mazim placerat facer possim assum.

Figure 7.2. Text flowed around a graphic image.

Further visual effects may be created by allowing text to flow over a graphic image. Provided the contrast of the image is reduced so that it does not interfere with the readability of the text itself, subtle effects can be produced in this way. In some cases, the position of an illustration in a publication is dictated by the design of the page layout and the amount of text to be included. When paginating a publication, the fitting of an illustration at a particular location may be subject to the space left on the page at that point. The importance of the illustration is another factor, especially if there is only one illustration to help identify the subject matter of the text which accompanies it. It may be helpful to the reader if the illustration appears either near the beginning or in front of the text, rather like the way a photograph in a newspaper is associated with the headline, rather than being slotted in somewhere else in the story.

Some documents are designed so that illustrations always appear in a particular location. For example, in a technical manual, illustrations might be reserved for right-hand facing pages and always appear in the same position on the page so that a reader can easily see the illustration which accompanies the relevant passage of text. Alternatively, the illustration may appear directly below the paragraph that makes reference to it, or to which it is relevant. Some desktop publishing systems allow illustrations to be associated with a paragraph of text (sometimes referred to as 'anchored' illustrations). The space occupied by the illustration is linked to a particular paragraph, or perhaps a heading, so that, when the publication is edited, should this result in a change in position of the text, the illustration that is associated with it moves too. The position of the illustration is always dependent on the paragraph's own position, but difficulties may arise when the paragraph moves to a position where there is no room for the illustration, either above or below the text. When this occurs the text may need to be cut and pasted so that the illustration can be included.

Some graphics, such as logos, may be required on every page of a document. Most desktop publishing systems permit 'master' items such as these to be specified, either as part of a style sheet or template, or as a special repeating item, so that the graphic appears in the same place on every page. One example where this might be required is the repetition of a company logo in the bottom corner of a page. In large format manuals with wide pages, a large margin on one side of the paper is sometimes reserved for the illustrations. In any case, if an illustration is not positioned directly with the text, it may be helpful to the reader to make some reference to it, and this is discussed next.

REFERENCING ILLUSTRATIONS

Whether an illustration is to be referenced because it is not near the relevant text, or simply because it is dictated by house style, all text references to illustrations should be consistent throughout the publication. It is usually best to adopt some kind of convention about illustration references. For example, one may use figure numbers, like this book, where the figure number relates to both the chapter number and the number of the figure within the chapter. One may simply choose to refer to 'the following illustration' or the 'illustration opposite'. If figures are used, the author must decide whether to refer to them as 'Figure x' or 'Fig. x' or 'fig x' etc, and then stick to the chosen style throughout the publication.

Whether an in-text reference is required at all depends upon how illustrations are used. If an illustration is not referred to in the text (perhaps because it is simply for enhancing the presentation), the use of figure numbers is largely irrelevant. If the illustration is self-explanatory, or carries its own brief description, again, an in-text reference is unlikely to be required. Should several illustrations be required on a page with a single paragraph of text referring to them, the separate pictures may be referenced with '(a)', '(b)', '(c)', etc., or they may be identified by their position on the page (such as 'top left', 'bottom right', etc.).

In some texts, such as technical books or training documents, where many illustrations are used, it can be useful to make the in-text reference to the illustration stand out from the rest of the paragraph, perhaps using a bold typeface. This will help the reader find the relevant text for an illustration quickly. In such cases, the illustration is the key to the subject of interest to the reader. Other means that may be used, apart from bold text, are the use of italics or capital letters.

CAPTIONS

The caption on an illustration may be used for a number of reasons. For example, it may simply carry a figure number, so that the reader must refer to the text to find out more about the illustration. It may carry some title or brief description about the illustrated subject as well as a reference, or may simply be a brief explanation of the illustration which excludes the need for any further explanation in the text.

The position of a caption may depend on whether it is also used as a heading (in which case, it may be positioned above the illustration) or on how much text is around the illustration (you may be able to fit the caption alongside).

If an illustration includes text, such as a graph which carries its own axis headings, etc., the author may choose to use a typeface, style and size that does not conflict with the illustration's own text, as this may cause the caption to become lost in the illustration. Alternatively, the illustration can be boxed to make a visual separation of the elements from the rest of the document's text as is illustrated in Figure 7.2 earlier.

SUMMARY

1. Graphics are often an important component within technical documentation. They may be in the form of circuit diagrams, engineering drawings, flow charts, graphs, screen illustrations or simply diagrams used to help illustrate a particular concept more clearly that the written text can provide alone.

2. There is a considerable choice of graphics application software available for both personal computers and more complex dedicated CAD/CAM systems. Libraries of graphics images and icons (known as clip art) are available on disk, CD-ROM, bulletin boards and other storage media providing a ready-made selection of symbols, diagrams, drawings, etc., that can be used within publications.

3. Graphic image files conform to one of a variety of file formats, most of which can be recognised by and imported into word processing and desktop publishing documents. The different file formats are suited to particular types of graphic image; for example, not all file formats are capable of storing colour images, some use file compression to help in handling complex and detailed images, etc.

4. Images included in word processing or desktop publishing documents can usually be manipulated by sizing and cropping. Depending upon the file format, some images can be altered so that contrast, brightness and colour attributes can be suited to the output device or to create a particular effect. Text may be flowed around graphic images which can be attached to a page or anchored to paragraphs, headings or captions.

5. The author should ensure that graphics included in a technical publication are properly referenced, especially if referred to within the text. The referencing and captioning system chosen should be consistent throughout the publication.

FURTHER READING

Austin, M. *Technical Writing and Publication Techniques* (ISTC, 1987).

Kay *Graphic File Formats* (Windcrest, 1993).

Omura, G. *Learn CAD Now* (Microsoft, 1992).

8 On-line Documentation

Although it seems likely that printed documentation will remain necessary in many areas of technical publications in the near future, documentation for the computer industry is moving ever closer to a paperless concept, particularly with the growth of multimedia systems that integrate video and sound. Because of its very nature, on-line documentation requires a screen on which to display and this restricts it for the most part to computer applications. Help text for software packages is the most common form of on-line documentation. Others include training and tutorial applications, on-screen demonstrations, CD-ROM-based animated displays, hypertext (interactive help text), video and sound.

HELP SYSTEMS

Help text systems for software applications comprise full or part screen displays, usually retrieved by a single keystroke, such as function key, or a key combination. Text may be displayed to provide an overall guide to the application — almost like an on-screen manual — through which the user can browse to find relevant information or may be restricted to covering a single topic relevant to the process being performed. This latter case is known as providing 'context sensitive' help.

Facilities for creating on-line documentation are numerous and varied as they are so often developed as part of the application they are intended to support. The main exception to this is in industry standard operating environments, like Microsoft Windows, in which all applications tend to provide help text that operates in the same manner and the tools for editing and generating the help information can be applied to any application working in that operating environment.

Help systems require a structured approach to providing information and provide an added challenge to the author in that they often have to communicate effectively in as few a words as possible. This could be argued as the same goal for those writing printing documentation, the help text author may not have the benefit of using the typographical or illustrative elements on a small screen display as can be provided on a printed page. Nevertheless, the rapid development of more powerful and efficient personal computers and processors means that the tools available to those involved in on-line documentation will be ever more effective.

Some courses specialise in teaching the techniques of on-line documentation. There is no doubt that the author requires a more concentrated effort in condensing useful information into small windows of communication, especially given that cross-referencing between topics and related information may be impossible (though see details about hypertext systems below).

HYPERTEXT AND MULTIMEDIA SYSTEMS

Hypertext systems provide a mixture of on-line help and tutorial information in the via an interactive interface. This simply means that the operator can choose to move around the help system on a key word or topic basis in order to find out more about a particular piece of information. Some help systems come close to be hypertext systems themselves. In the Microsoft Windows operating environment, as mentioned previously, context-sensitive help can be compiled with what are called 'hot keys'. These are words or phrases or even graphic that can act as a trigger to retrieve more information about a related topic. The key words or phrases that form the hot keys may be displayed in a different colour so that the user knows that by clicking on a hot key with the mouse pointer, another window of help information may be displayed offering another level of details about the topic concerned. Figure 8.1 shows an example of a help screen in the Microsoft Windows environment. The words with dashed underlining are the hot key phrases and words.

Hypertext systems can become an integrated part of an application. They may result in a series of actions being taken when the user chooses a particular item or menu choice which may include audiovisual sequences, further levels of help information or may even cause the application to perform a particular function. Hypertext systems can also be developed as independent applications in their own right, offering information about a particular subject and may allow the user to browse through the system in much the same way as someone might look up information in an encyclopedia. With the advent of multimedia personal computers, such applications may be purchased on CD-ROM or other storage device, replacing more conventional learning media. In a commercial environment, hypertext systems may be used to replace printed documentation or as a tutorial or demonstration for a software application that may be operated without the need for the source application to exist.

Another kind of hypertext system is the 'expert system'. This is what is currently referred to as artificial intelligence (AI). There are some software systems that offer this form of on-line documentation in association with an application. Essentially it is a

ON-LINE DOCUMENTATION 115

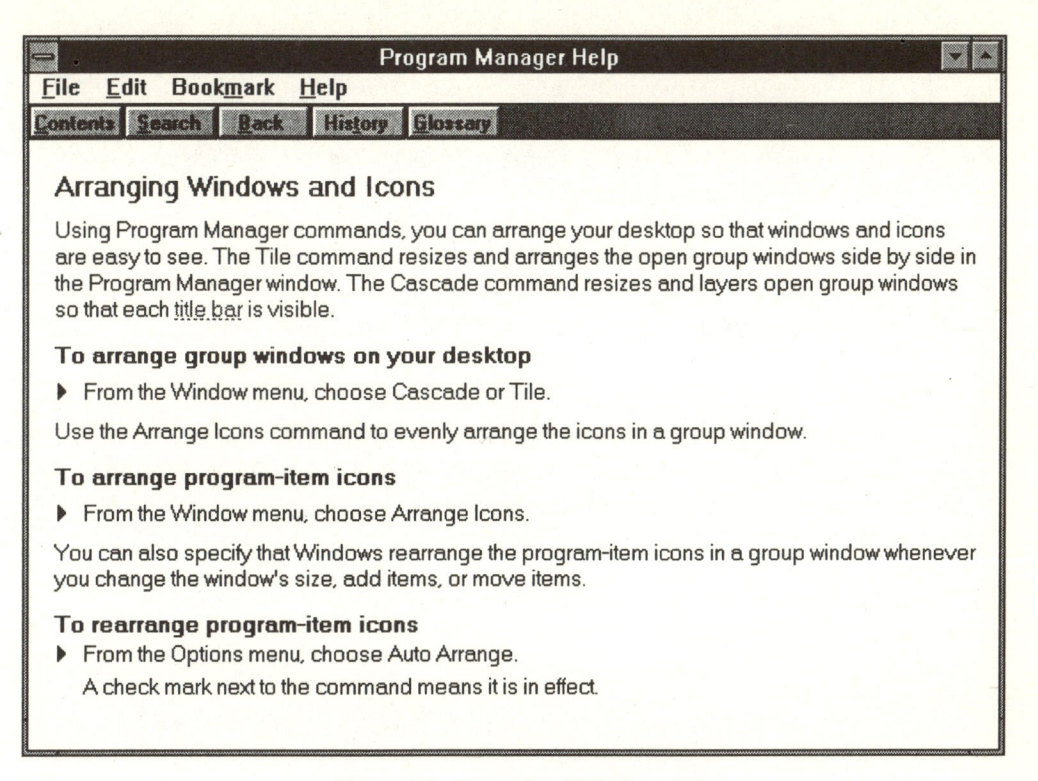

Figure 8.1. Example of Windows help.

database of information that provides answers to questions posed by the user. Those questions are inevitably structured and in some cases the user may be given a limited choice of options. Using a structured and logical design, a solution, an action or simply more information can be provided in response to a series of queries. It is rather 'twenty questions', in which the user gradually arrives at the information or action required by responding to progressively more detailed queries. This makes expert systems ideally suited to handling troubleshooting documentation. Rather like visiting a GP when one is ill, the expert system can diagnose a problem on the basis of answers to a structured set of questions. Expert systems, like hypertext system of help, can be used in place of or alongside traditional documentation. The development of such systems has been going on for some time, though, at the time of writing this book, they have still to realise their promised potential.

Since the subject of multimedia can cover a significant range of additional specialist skills through graphics, video imaging and processing, sound recording and mixing as well as text-based authorship, this book can only introduce the concept of this form of technical documentation. Inevitably, this form of communication will become more important in future years as the need for printed documentation is eroded in the computing industry. However, printed documentation remains an important component in this field at present as well as being an essential accompaniment to hardware and systems that cannot provide a means of interactive communication through the product itself. No doubt this will change in the future too and even the humble electric toaster may offer instructions on its use by a built-in sythesized speech processor.

The role of the technical author is likely to change in future years as it becomes more specialised and as printed matter is gradually replaced by multimedia forms of communication. It is therefore important for those involved in technical documentation to be aware of a keep an interest in developments in on-line documentation.

SUMMARY

1. On-line documentation is information displayed, usually on a computer screen, which informs or instructs the user of a system or application. On-line documentation may be presented in many forms from error messages, warnings, help, tutorial information, advice, etc.

2. The approach to designing and writing on-line documentation differs from printed text as the style, appearance and extent of the information is often restricted or limited.

3. Some on-line documentation provides an interactive interface for the user. These systems enable the user to browse through a database of help or tutorial information either by choosing menu options, responding to queries or selecting key words or phrases that provide access to information at another level. Systems that fall into this category include hypertext and expert systems.

9 Publications Management

The role of the publications manager is a diverse one and encompasses the responsibilities of author management and print production control. If one were to summarise some of the main duties of the publications manager they may include the following:

— the recruitment and development of technical authors;

— the initiation and control of documentation schedules;

— overseeing the design and layout of documentation jobs;

— budget control;

— print buying;

— progress chasing.

THE PUBLICATIONS MANAGER

The role of the publications manager may well include many of the responsibilities associated with that of the author-manager. As with any other manager in an organisation, the publications manager has to undertake to successfully control the workings of the department and to develop and use the resources within it to the benefit of the organisation and its customers.

The publications manager is responsible for the functioning of the publications department at all times. He or she must ensure that all staff in the publications unit are properly trained for the tasks expected of them and ensure someone is responsible for each task or project at appropriate levels. It is the responsibility of the manager to set up and maintain a strong team, to ensure that standards and quality are maintained, and to provide staff under his or her control with the appropriate technical training, motivation and sense of responsibility. In addition, work and responsibilities should be delegated, but not abdicated. The role of the publications manager should also include the following objectives:

— to set aims and goals for the authors and supporting staff;
— to subdivide and allocate the documentation workload to the appropriate individuals;
— to monitor the progress of the staff and current workload, checking against plans;
— to develop people, or to manage them in a way that enables them to develop themselves.

All of the above could be a list of responsibilities for the manager of any department. They are important points of management and managing people in a publications environment has few unique characteristics which set it apart from the management of a sales department, administration department, or whatever.

PLANNING DOCUMENTATION PRODUCTION

The publications manager will be involved in planning the content of the documentation in advance of any writing, so that a structure can be referred to throughout the process. Although the task may be delegated to one or more authors, the publications manager should, at least, approve a document's structure at the start of a project and ensure that the content and approach meets all necessary criteria. If documentation is to be produced to a specification or standard, the publications manager should ensure that all concerned are aware of the details and have copies of the specifications that should be followed.

Planning the structure of a document for external readership requires input where possible from the reading audience. Their input can help the publications manager to assess the required content of a document before planning tasks begin. There are various ways in which the document's structure can be planned. One way is to first attempt to produce the list of contents for the main chapters. This will help put into logical order the main categories of text. These main sections can then be broken down into subdivisions, followed by synopsis of the content of all sections. Planning will also encompass illustration requirements, layout, design and printing.

Thorough planning of all factors before the documentation cycle begins will keep the project smooth running. For example, the publications manager needs to know how a job is to be presented when finished, before piecing together the structure of the document in detail; there is little point in planning a document that relies heavily on photographic illustrations if neither the budget for the job, nor the specification permits the use of such illustration material.

A more experienced author or publications manager will generally need to spend less time planning the documentation, and as house styles are specified and similar documentation jobs are processed, the planning tasks become less time consuming as requirements in this direction become well defined.

RECRUITING AND AUTHOR MANAGEMENT

The responsibility of any manager who recruits staff for the organisation should not be taken lightly. As a management duty, the recruiting of candidates for vacancies is probably one of the most important tasks to be undertaken. It is difficult to identify what characteristics or skills make a good technical author. The chapter on the author's role earlier in this book discussed some of these characteristics, but decisions cannot be based on lists of qualities alone. For example, the publications manager needs to consider whether the personality of the candidate will fit into an existing publications team.

Qualifications that are relevant are hard to identify. It may be that an author candidate has attended some formal training or short courses on documentation or technical writing, but one may feel that it is more important for them to be qualified in the subject area about which they will write.

Experience will be of great value in the successful selection of good authors; experience in both the manager making the selection and of the candidates themselves. However, the publications manager should not overlook the possibility of taking on trainee authors as this can reap long-term rewards as the authors can be developed within the most appropriate environment for the organisation's documentation requirements and not influenced or distracted by previous irrelevant experience or learning.

Once in the team, the newly recruited author must be developed to the fullest potential. This may require on-going training, and the publications manager should take the trouble to provide a training plan so that the author can improve on current abilities and gain a wider experience of method. An author can often be an isolated individual, getting little chance to meet colleagues in the industry in the same line of work. Seminars and training courses go some way to at least providing that opportunity on a short term, but it is a good idea for the author (and indeed the publications manager) to get to know how others cope with the job.

Typically, the progression for authors is into areas of author-manager or publications management, but with technology changing and developing rapidly, the possibilities for structuring the duties and responsibilities can be more varied. For example, allowing authors to learn to handle basic design principles and then giving them the opportunity to try these out on documents using DTP can be helpful in improving job satisfaction. Authors will be naturally curious about any new publishing technology, and spreading control over technology that may be available in-house provides ways of adding variety and interest to the task of technical writing.

Publications managers should counsel their authors, just as any other managers would do to assess the development and interests of the individual authors under their control.

Not all technical authors want to stay technical authors and one can become complacent about this fact. This may be because the job of technical writing is difficult to get into in the first place, so one may wrongly assume that an author who has struggled to get into the profession is set for life.

PRINT BUYING

This is a task that requires experience as well as basic training. The publications manager will need skills that involve being able to plan, prepare, budget, order and progress-chase publications from conception to completion. If the documentation to be produced involves the use of external resources for writing, typesetting, illustration or any other artwork needed, the print buying role will include the control of such services.

How detailed the role of print buyer will be depends on the complexity of the jobs concerned. If documentation is word processed, photocopied and collated in small quantities, there will be little responsibility for the publications manager in this particular area of activity. However, many publications demand more sophisticated methods of publication and may involve litho printing, binding, packaging and other forms of presentation.

Print buying is defined as the management of all the resources available to meet an organisation's print needs. To do this effectively, the publications manager must not only be knowledgeable about the processes already employed in production of existing documentation, but also other internal and external processes and materials at the disposal of the company. Doing so ensures that decisions can be made effectively and wisely. Print processes are heavily technology oriented. Like most forms of technology, therefore, new developments are introduced regularly over short timescales which can make the task of the publications manager who must keep up with the technology available, that much more difficult and time-consuming.

Only a few years ago, the concept of using full-colour processes in technical documentation would have been considered too expensive. Yet with the introduction of desktop colour photocopying and laser or ink-jet printing, with colour scanners and appropriate software, full-colour documentation can be produced in-house at competitive prices, especially where print quantities of finished documents is relatively low. Where the extent of a document or the quantities required are high and litho printing is required to cope with the demand, technical documentation, such as user manuals, may still be too expensive for anything more than a second spot-colour, but advancing print technology is likely to remove this barrier.

It can be quite a task to assimilate the knowledge required to both understand and make best use of the facilities and services available to a publications department. The publications manager may be involved in buying both services and materials, and there is a great deal to choose from in both categories. Some organisations prefer to employ the use of an agent to do the print buying, such as an advertising and/or marketing agency with print process knowledge. This has the advantage of easing the burden on the publications manager's job, being less time-consuming. The task becomes one of submitting requirements to the agency who then advises on the most appropriate production method and takes on the print buying role, and possibly even part of the artwork production of the documentation. Progress chasing for the job is then restricted to chasing the agent rather than a number of contractors, printers, finishers and other services involved. The disadvantages are lack of control on the part of the publications manager and the possibility of incurring greater expense.

When dealing with printers, the publications manager should take the time to visit their premises, and talk to them about how they do their job. They may not volunteer any background knowledge, but are usually quite happy to oblige if asked. The publications manager can pick up a lot of information about various aspect of printing by getting to know the printers involved with technical publications and for an inexperienced publications manager, this can be well worthwhile.

Once the publications manager has gained knowledge about what different processes are available and what the current technology offers, he or she has the fairly awesome task of understanding what types of job suit what processes. This involves being about to identify the appropriate materials and processes for a particular job. Some publications will be suited to production using paper printing plates, others may need processes that use web-offset presses using plastic or metal plates and so on. Once familiar with the differences, the manager will be able to make better judgements of how best to tackle the organisation's technical publication print requirements. It should not be overlooked that many printers will advise of the best processes for the job anyway. It is their job to know what process will be required for any particular job. However, the publications manager who accepts quotations from printers for various projects simply because they fit the print budget, may never get to appreciate what process is being used. In the light of the pressures of the job, it is easy to leave the responsibility to the printer alone, and not to get involved any further. If given a ceiling on the cost of a job, the printer will always choose paper, other materials, and a process on the basis of the available budget, rather than what necessarily suits the job best. Unless the publications manager takes time to understand the options available, he or she will be in no position to change this state of affairs.

Print Specifications

The inexperienced publications manager may find it useful to draw up print specifications for jobs that have to be printed. The specification should include details of the format used, e.g. A4 orA5, whether the printer is being provided with camera-ready artwork (phototypeset or laser printed), finishing requirements, etc. Figure 9.1 shows an example print specification for a user manual for a software product.

Once defined, the specification can be used to obtain quotations from more than one printer, to help assess the cost of printing and in an effort to reduce costs where possible. The specification becomes as detailed as the complexity of the processes involved, so

Acme Publications Department
Job Processing Schedule & Print Specification

Job No. _____   Job Name _____

File Opened ___/___/___   Due Completion Date ___/___/___

Details _____
_____
_____

Print Specification _____ Qty Req. ____

Type _____ Colour(s) _____

Style _____ Print Finish _____ Paper _____

Illustrations _____ Throw-outs _____ Binding _____

Packaging _____

Special Requirements _____
_____

Status Summary:	*Due Date*	*Actual Completed*	*By Whom*
Familiarisation	___/___/___	___/___/___	_____
Documentation	___/___/___	___/___/___	_____
Word Processing	___/___/___	___/___/___	_____
Copy Proof	___/___/___	___/___/___	_____
Illustration	___/___/___	___/___/___	_____
Page Proof	___/___/___	___/___/___	_____
Amendments	___/___/___	___/___/___	_____
Artwork to Printer	___/___/___	___/___/___	_____
Delivery	___/___/___	___/___/___	_____

Figure 9.1. Sample print specification.

for a document that is bound in a ring-binder and slip case or packaged in a box, this should include details of the materials used for packaging, the use of colour, and identify who is responsible for the origination of artwork used. This latter point is important. Estimates from printers may, unless specified beforehand, only quote the print costs, leaving any artwork processes as a variable additional cost. It is therefore important to identify who is responsible for handling artwork origination. If the printer can be given complete camera-ready artwork, then the printer may only need to cost the imposition, platemaking, printing, collating and finishing elements. If artwork has to be prepared by the printer, the overall and unit costs can be affected substantially depending upon their facilities, as the printer may agree to take on artwork production, but contract it out. It pays the publications manager to know what facilities the printers.

The print schedule includes details of printing timescales, though it may not be possible to specify this, except in individual cases at the time they occur. However, if very short timescales are involved, then it is important that this requirement is identified to the printer. Print presses do not have limitless capacities and are usually tightly scheduled to ensure that presses are running continuously for the sake of cost effectiveness. If that process is to be interrupted, either by demanding very quick turnarounds or by continually putting off submission of copy or artwork because one's own schedule is changing rapidly, then this may cost extra. While printers may be able to 'pull a few strings' when something is wanted quickly, they may need to employ overtime rates, and the job will be correspondingly more expensive. These points are made because in many areas in which technical documentation is produced, particularly where it accompanies a new product to be launched on the market, print timescales are invariably short as documentation is all too often one of the last stages of the development cycle.

Supplier Selection

Selecting the supplier for print jobs is not an easy task to define using any standard criteria. Much will depend upon individual requirements. Locality may be important, for example, though many print jobs for technical publications are adequately handled in more remote sites.

It is not unusual for publications managers to experience 'cold call' contact from a printer's representative looking to take on the organisation's technical documentation print requirements. Beware of those printers who claim to specialise in printing technical publications. Some genuinely provide services that can assist an organisation in technical publications by offering facilities for short print runs, specialist print finishing and possibly distribution. There are, however, some printers who offer their services as specialists in this field, simply to attract potential customers with technical publications, but in fact contract out some of the processes mentioned and, as a result, can be less price-competitive than even non-specialist printers.

If the printing requirements of the technical documentation involve straightforward litho printing and simple finishing such as perfect binding, there may be no advantage in using so-called 'specialist' printers at all. It is more important to find a printer who has the most appropriate equipment for the type and size of job. For example, a printer with a single-colour press may be able to produce two-colour work by putting a job through twice with different inks for each run, but this may not be as cost effective as a printer with a two-colour press.

IN-HOUSE PUBLICATION SYSTEMS

Generally speaking, the more an organisation can afford to handle print production in-house, the more control, both productive and economic, one has over printing requirements. The production of artwork in-house is certainly far more widespread since the development of desktop publishing systems and is certainly very affordable at the entry level. However, as will any field of specialisation, such facilities can only be used properly by those trained to do so.

Good management is required to ensure that quality is maintained and unless the appropriate skill levels are as available in-house as the equipment, the publications department will achieved only a limited variety of output and relatively low quality standards.

In-house publication systems must be justified both in economical and quality terms and need to be an effective alternative to using external sources. A company may be able to afford desktop filmsetting technology to produce optimum quality artwork for documentation, but if the throughput time is slow compared to a more powerful external bureau service, the benefits of cheap and controllable typesetting may be outweighed by the extended lead time which may cause unacceptable delays in delivering the finished job.

The publications manager need to be conscious of the limitations of in-house systems as well as their benefits. External sources that provide specialist services, whether they be technical illustration, distribution, design, etc., may be better skilled and equipped to carry out particular parts of the publication process that in-house staff with similar, but ultimately less flexible equipment and it may not be possible to cost-justify the purchase of very specialist prepress facilities.

The publications manager may be responsible for the acquisition of equipment and software used within the publications department. The more complex and widespread the in-house responsibilities for technical publications become, the greater the responsibility on the publications manager to choose wisely. A large organisation producing a significant amount of technical documentation may be able to justify an in-house printing and finishing department. Such departments need to be supervised by people skilled and experienced in this particular field and publications managers looking to develop such installations within their own organisation may need professional advice to make up for any shortfall in their own training or experience in the printing industry, especially when procuring printing and print finishing equipment.

The level of in-house publication facilities that an organisation can support will depend on balancing the benefits of time, cost, flexibility and available skills against the short- and long-term strategies for technical publications. It may not be possible to justify in-house phototypesetting, or technical illustration skills in the short term, but, after taking into consideration the rate at which technology changes and generally becomes cheaper over relatively short periods of time, may well be a cost-effective solution in the long term.

COSTING AND SCHEDULING

These two areas of responsibility are among the most important of the publications manager after the recruitment of skilled personnel within the publications department.

Costing

The costing of documentation projects is necessary to enable budget variance control and to help analyse performance and expenses, reduce costs and improve financial efficiency. It is also necessary for invoicing purposes should your documentation service be saleable. To cost a documentation job, you need to break down the project into the components or cost types to an appropriate level. For example, you may consider that, for your purposes, you need only split costs into four categories:

Cost to Write + Cost to Review + Cost to Produce + Cost to Print

Within each of the above categories, you need to identify the associated cost types. 'Cost to Write' will certainly include the cost of the author's writing time, but may also include the cost to word process, the cost of the author's time in learning the subject matter, etc.

'Cost to Review' will identify the costs of proofing and amending at all stages of draft and final copy production, but may exclude cost of printing proof reviews which could be included in one of the other two categories.

'Cost to Produce' could include all artwork preparation costs, illustration costs, etc., and all categories should include their respective expenses and material cost requirements.

There will be a good deal of personal preference to costing methods and there are few basic principles that deserve any amount of detailed explanation, but the following paragraph provide some simple guidelines.

Where possible, you should try to allocate budgets for both individual jobs and the cost types. So, for example, a list of job budgets may look like this:

126   TECHNICAL DOCUMENTATION

Job No.	Description	Budget £
J001	System Specifications	6,500
J002	User Handbook	8,000
J003	Installations Guide	2,000
J004	Technical Fault Guide	1,200
	Subtotal — Product × Documentation	17,700

etc.

The budgets for the cost types is specified in a similar manner, e.g.:

Cost Code	Cost Type	Budget £
W01	Author 1 — Standard Rate	9,000
W02	Author 2 — Standard Rate	9,000
W03	Author Manager — Standard Rate	12,000
	Subtotal Authors	30,000
WP1	Word Processor — Standard Rate	6,500
PC1	Photocompositor	15,000
	Subtotal Pre-press	21,500
M01	Materials — Paper	2,000
M02	Materials — Toner/Chemicals	1,500
M02	Materials — Binding/Packaging	3,500
	Subtotal — Material Costs	7,000
OH1	Overheads — Equipment Maintenance	1,000
OH2	Overheads — Equipment Depreciation	4,000

etc.

In the above examples, the job budgets are allocated to each job as a specification is drawn up in the early stages. The cost type budgets are allocated on an annual basis and then costs are booked to the individual jobs as they arise. The annual budgets can be broken down into period budgets, such as monthly and monthly variance reports can then be produced. In the case of time costs, these will be booked against the job on an hourly or daily rate basis, then totalled for budget comparison. The materials costs are booked against the jobs to which they apply, but overheads are not associated with the individual jobs, applying to the department costs will vary depending upon the detail of cost control you want associated with any given job. In the above example, you may consider costs such as paper and printer ribbons as overhead costs since you do not need or want to break down these costs and allocate them to individual jobs.

Figure 9.2 shows an example budget analysis report for a series of documentation

```
10.12.93                    Job Cost Variance Report                      Page 1

Job Nos.Description                   Budget      Total      Variance    % Var      &
                                                  Cost                              Compl.

UM-0031   User Manual Printer V1.2    2000       1946.25       53.75      2.69     0.97

UM-0042   User Manual Alpha Product   4000       4101.00     -101.00     -2.53     1.03

UM-0043   User Manual - Centurion     4000       1009.10     2990.90     74.77     0.25

MB-1014   Product Brochure           12000      10989.56     1010.44      8.42     0.92

PL-0030   Price List                  1500       1585.00      -85.00     -5.67     1.06

PC-8010   Parts Catalogue             5500       6010.00     -510.00     -9.27     1.09

TS-0113   Printer Manual/Tech. Spec.   500        395.50      104.50     20.90     0.79
                                     -----------------------------------------------------
Totals                               29500      26036.41     3463.59     11.74     0.88
                                     -----------------------------------------------------
```

Figure 9.2. Example budget analysis report.

jobs. The variance information is useful to monitor expenditure and to enable future budgeting to be based on real costs will more accuracy.

When costing resource time, all relevant costs should be included for each resource. For an independent cost analysis as a publications department, it will be necessary to include overhead costs that relate to the publication department as a whole as well as to the individuals employed within. For example, an author's salary may be £18,000 per annum, but National Insurance contributions paid by the company should be added to this, since this is part of the expense of employing the author. So, if the prevailing rate was, say 10%, this would add £1,800 to the overall author costs. If the author receives other valued benefits, these must be included also. These benefits may include pension, car, expenses, etc. If a resource uses expensive equipment on a permanent basis, the cost of depreciation of the equipment used should also be included as part of the total resource cost. In these cases, the cost analysis is geared towards costing of the whole department, and the job costs that result should take on the appropriate weighting of all departments overheads. Such a costing method is essential to technical publication units or subcontracting authors who charge out their time for producing documentation for clients. They will also have to uplift the costs, probably by a percentage, to make a profit on top. However, for internal costing purposes, all job costs should remain 'at cost' only. A example of a formula for working out the rate cost for an author resource, may look something like this:

(Author Salary + N.I. + Benefits + Expenses + Equipment Depreciation) ÷ (period of costing)

Notice that all items in the formula would be annual costs, which, when divided by the appropriate costing period, would give a cost per month, week, day, hour, etc., depending upon the period concerned. For example, if the author works 230 days per annum, the costing period of 230 would yield the cost per working day. If the author spends only a certain percentage of time actually writing documentation, perhaps because it is a secondary activity, it would be unfair to cost the full daily rate on a job. Therefore the costing period would need to be adjusted accordingly.

The publications manager needs to arrive at a costing method suitable for the particular requirements and circumstances within the organisation. It may be, for example, that approximate costs are sufficient and that individual resource rates are do not need to be calculated so precisely. The important point is that whatever costing method is used, it should be applied in the same way to all similar jobs to ensure effective cost management.

Timesheets should be used if accurate cost control of resource time spent on individual jobs is required. This will be of particular importance in a department of many resources which include authors, word processor operators, graphic artists, project managers, etc., all of whom may be working on different tasks at the same time. Timesheets are used to gather the data required to place costs against a job, by indicating the number of hours worked (including any fractions of an hour if necessary) on a given documentation project or task. The cost rate calculated for the resource can then be use to arrive at an overall cost of producing the particular job. Figure 9.3 shows an example timesheet.

```
13.12.93                    Timesheet            Page 1

Employee 4005         Timesheet for w/e 10/12/93

Job         Cost  Units  Rate   Value
Number      Code

UM-1021     A001  6.00   10.50  63.00  1st draft authoring        6/12
UM-1021     R001  2.00   7.80   15.60  Review draft              7/12
UM-1043     A001  4.00   10.50  42.00  Writing updates           8/12
TS-1132     I012  3.50   10.50  36.75  Illustrations             9/12
UM-1021     A001  5.75   10.50  60.38  2nd draft authoring      10/12
00-0000     UNPR  15.00                Unproductive time (training)
                  -----         ------
                  36.25         217.73
                  -----         ------
```

Figure 9.3. Example timesheet.

Like resource costs, materials which can be identified as being used on a specific job should be costed to that job. As values are accumulated against the various jobs in progress, the publications manager can at any point in time or over particular work periods, compare the cost of materials booked to jobs with the relevant budget for expenditure on such materials. The same applies to any specific overheads. If the department produces artwork in-house, such as phototypesetting or illustrations, the relevant running costs of the equipment will need to be booked to the relevant job(s). It may prove useful to work out a machine or facility rate, such as the use of a DTP system, so that work done in that particular area can be analysed through timesheets, just as personnel costs would. The machine or equipment rate can then include the associated costs of the system (e.g. toner cartridges for a laser printer, paper, maintenance contract costs, depreciation, etc.).

Apart from the costs of processing jobs internally, some technical publication jobs may require the use of some external services such as designers, technical illustrators, authors, typesetters, printer and so on. These costs will be easier to associate with individual jobs, since the publications manager will probably get quotations first. The same applies to purchases of materials such as paper, ring-binders, etc.

It can be a useful exercise to calculate the cost per page of each technical publication job. This may even be extended to an average cost per page of any technical publication, provided the types of job are similar. The cost per page is arrived at by dividing the unit cost of the job the number of pages of the finished publication. The unit cost of documentation can be calculated by taking the total job cost and dividing it by the number of finished units (e.g. the number of manuals printed). The unit cost will vary according to the number of units printed because the larger the print run, the lower will be the unit cost.

Where technical publication jobs tend to be similar, the cost per page can be used to calculate the approximate costs or budget for similar future jobs, though it relies on being able to estimate the extent of future publications. There are so many variables involved in calculating the cost of publications, including the complexity of the work involved, that it may not be practical to use the cost per page for anything other than analysis purposes. Publication managers that use external documentation services may find this exercise useful for comparison purposes for similar jobs produced by the same or competing bureau services to keep a check on the level of charging.

Scheduling

Planning and scheduling for publications is just as important for technical documentation production as it is for any other type of production process. It is not uncommon to find that the process of scheduling is left to the product development department of an organisation and that the documentation is simply appended to the development cycle. This may be more apparent in smaller enterprises, but less so in larger corporate companies or institutions where stricter control in budgeting and scheduling is generally applied.

The publications manager will be responsible for scheduling technical publications within the department. This process should begin by breaking down the various tasks into components of the publication cycle. This cycle may depend heavily on the development cycle of the product or system to which the documentation relates. In some cases, this may be subject to frequent and rapid changes. In such circumstances, the schedule needs to be flexible.

Figure 9.4 shows a publications cycle broken down into main components. This can apply to a wide range of technical publication jobs but does not necessarily relate to every stage of a given job. For example, a manual may require several proofing and editing cycles, perhaps involving many departments or individuals before final artwork can be produced. While in some cases, the entire process may be carried out by one person. It may not even be possible to assess how many proofing and editing processes will be required, especially if the documentation project is to be subcontracted out and the quality of each draft is unknown in advance. The overall flow from one stage to another is likely to remain constant; i.e. reviewing must follow first draft which must follow general design and conception, etc. The following is a summary of the main activities included in the example publications cycle:

General design concept — this covers the preparation work undertaken before any authorship begins. This may included an overall design for the documentation, page layout, packaging, distribution, quantity required, etc.

Authoring first draft — the first draft of a document may include other activities, such as the learning curve for a new product or system that is being documented, familiarisation with house style, word processing, etc. If the documentation project is complex and involved many components, it may be pertinent to draft only part of the documentation before carrying out the review process, so the first draft stage be extended beyond the first review stage.

DTP — in some cases, this may be carried out at the same time as the writing if the author uses desktop publishing software as a writing tool. If so, this stage of the publication cycle may be incorporated into the writing or drafting stages of the project. If word processing or desktop publishing is carried out as a separate process, there may be a need for separate reviewing stages to check the accuracy of typing as well as technical content.

Proofing and editing — as already mentioned, proofing may be involved at several stages and may be carried out by many different departments or individuals. Documentation may be included as part of a quality control procedure handed by a simple quality control department, or it may be reviewed by technical experts, sales and marketing departments, training staff, support staff and others who may be associated with or involved in the product or system to which the documentation relates. Editing may be under separate control to the author of the draft.

Illustration — like the DTP element of a publication, illustration work may be carried out by the author while drafting the documentation or may require specialist engineering designers or technical illustrators. Illustration material itself may be subject to a proofing and approval processes.

Artwork preparation — this is the time allocated for the preparation of the final copy that is used for production, which may be output from a laser printer, phototypesetting or filmsetting in the case of printed documentation. The final stage of a technical publication, however, may also be some other medium, such as video, disk-based help, CD-ROM-based animated audiovisual, etc.

Printing and finishing — as with artwork preparation, this can vary according to the nature of the technical publication. If this part of the process is carried out by an external service, such as a printer, the scheduling of this tasks will be the responsibility of the printer/finisher concerned.

Some tasks in a schedule cannot be carried out unless preceding tasks have been completed. Clearly, printing cannot commence until the artwork has been prepared and artwork cannot be prepared until the document is written. Such tasks comprise the 'critical path' events; i.e. they are events that are wholly dependent on the completion of others. Non-critical events are those that can be completed in parallel with other activities. For example, it may be possible for one part of a user manual to be reviewed and edited while another part is being drafted.

To help in the process of project scheduling, the publications manager may wish to utilise one of the many project management software applications that can be purchased for use on personal computers. Such applications enable a schedule to be manipulated by controlling particular event dates to provide an immediate recalculation of completion dates which is particularly useful where a project is subject to changing circumstances or variable availability of resources.

QUALITY CONTROL

Quality control is just as important in the area of documentation as it is for the product or system for which the documentation is intended. However, it is a process which is either overlooked or omitted due to the lack of time available. Lack of time is a frequent problem in documentation production since it is often a task left until the end of a product or system development cycle. This can usually be overcome by better planning at the outset.

Documentation should be checked carefully to ensure that the required standard is achieved before going to print. The following is a breakdown of general areas that should be included in any quality control programme for documentation, though additional areas may be required according to the subject matter, intended readership and method of production:

— accuracy of technical content;

— conformity to any applicable standards;

— conformity to house styles;

— grammatical correctness;

— consistency of writing style and terminology;

— correct use, positioning and referencing of illustration material;

— accuracy of contents, section numbering, indexing and cross-referencing;

— accuracy and consistency of typographical presentation.

A publications manager should be aware of other documentation within the same field of interest as the organisation in which he or she works. Managers should make it their business to know about this, especially if they are providing documentation for products that are sold in a competitive environment. This involves an evaluation process of both the organisation's own documentation as well as that published by others. It involves looking for both the good and bad qualities.

The publications manager should read competitive documentation for style, tone, consistency and general structure. It is very easy to criticise other documentation and use other examples of documentation as simply an ego-boosting trip for one's own material. Authors and publications managers are naturally proud of their own work and assume that it is better than any documentation in a competitive environment. This leads to a cynical approach to evaluation of other documentation. Being objective is the key as well as being able to recognise those elements of other documentation that is better that one's own.

While evaluating documentation, whether competitive or 'home-grown', the publications manager should check for the accessibility of information contained in it from a reader's point of view. This involves 'testing' the documentation in a 'user' environment, to see if the level of indexing, cross-referencing and section order is satisfactory. These matters can be overlooked when one concentrate's on the accuracy of content alone, and even when checking for completeness. Good user documentation should guide the reader to and around all the appropriate sections of the text to perform its primary task of communicating information. Once the qualities of the content of competitive documentation has been assessed, comparisons should be made of the more aesthetic attributes such as page style, typography, use of graphics, colour and materials, such as paper, binding and general print quality.

Another way of evaluating documentation is to take the time to listen to, and gather up, opinions from readers. This can be achieved in a number of ways. For example, in the case of a user manual, one can arrange to have the documentation 'site tested' with the product before general release. The 'test site' can become part of the proofing cycle of the documentation project. The only problem is that a user site testing some software or hardware product may have limited interest in improving the documentation, so without real incentive, they are unlikely to provide comprehensive feedback, and should not be relied upon as a quality assurance exercise alone; it is better that the documentation is tested as part of the quality assurance procedures for the product to which it relates.

A method employed by some organisations is to provide a market survey system within the documentation itself to obtain feedback from readers. However, for this to work, the kind of information required in response of, say, a questionnaire, must be planned carefully. Some products and/or their documentation include reply-paid cards on which the user or purchaser can comment on the quality of the product and its components, including documentation. This approach can serve a number of purposes. Market survey information can be gathered by asking questions that identify the type of user or purchaser. The reply can be used as a form of product registration for on-going support, maintenance or after-sales services. At the same time, the respondent can be encouraged to provide feedback on the product and its documentation. Any part of a survey intended to provide feedback on documentation must be structured with care to avoid ambiguities and irrelevant responses. A simple multichoice question such as 'do you think the documentation was excellent, very good, good, satisfactory, poor, very poor' may not yield any useful information unless the reason for the response given can be qualified. For example, a user may have referred to the documentation only once in order to solve a particular problem. If a solution was not found, the opinion of the user may be that the documentation is poor. Such as response on a survey form would not yield a fair judgement on the quality of the documentation as a whole, which may be good in many other ways, including general content, structure and design.

One should recognise that users will be reluctant to complete detailed questionnaires simply for the benefit of the manufacturer. Offering incentives can be one way of ensuring a reasonable response (such as a chance to win something extra if a reply card is included in a draw), or it may be made compulsory as part of the registration of the product (as may be the case with many computer software products). It is important to keep any questionnaire both simple and fair. A computer software company once included the following item on a reply card for feedback on its documentation: 'Please indicate if you have found any errors in the documentation, stating briefly the error and the page number'. Only one form was returned that included a response to this prompt, and it stated, quite rightly, that 'checking for errors is your job, not ours!'. A sober lesson to any publications manager or author, as such requests are not only an invitation for trouble, they imply that errors will be found, possibly undermining the confidence of the reader.

Finally, users of documentation will help the evaluation process by providing feedback that is not necessarily asked for. This is generally in the form of complaints, given that few will inform the publishers that everything is okay. There are a few who take the time to praise documentation, but this is rare. However, criticism can be more useful, since it is on this that the publications manager should take appropriate remedial action.

SUBCONTRACTING

The subcontracting of technical authorship to freelance authors or documentation houses may be an important and useful consideration for the publications manager. There are a number of reasons why an organisation may wish to employ the services of external technical authors including:

(a) high workload;

(b) specialist skills required;

(c) timescale (urgency);

(d) cost effectiveness;

(e) infrequency of technical publication production.

These reasons for using outside sources are important as the publications manager should always consider the most effective way to produce the quality of documentation required. In the first reason given, it is common for organisations to employ the services of freelance authors or documentation houses when the workload is suddenly beyond the means of the in-house facilities and does not justify the increase of the in-house personnel on a long-term basis. Even if the subcontractor's costs are higher than an employed technical author on an hourly or daily basis, the overall cost of the job may still be cheaper than the cost of recruiting, training and keeping a full-time author.

Specialist skills are another common reason for using subcontracted authors. A particular documentation job may require the experience of a technical author in a particular field or discipline not covered by those individuals available in-house. As a result, it may be more cost-effective to employ a freelance author than to train an existing or new employee in the skills required.

The urgency of a job is one aspect that is not immediately apparent as a benefit to using external services. However, it is a fact that employees are unlikely to work at the same speed or level of commitment as a freelance or subcontracted writing service where the completion of a task on time is more directly concerned with their livelihood. A freelance author may be prepared to work long hours in order to obtain or retain business with a particular client and will feel more under pressure to perform well and on time than a employee. An employed technical author may feel more safe in their job and is almost certainly able to delegate responsibility for ineffective unproductive time on other departments or individuals within the same organisation. The obligations on an author paid on a job-by-job basis to perform effectively and efficiently are considerably more apparent than an employee who is paid at the end of each month regardless of productivity.

This latter point is partly connected with another advantage — cost effectiveness. Subcontracted authors may be more expensive that in-house authors; especially on an hourly- or daily-rate basis. However, few publications managers actually measure the cost effectiveness of their own employees' productivity. Time lost due to illness, holiday and general job complacency can be costly to an organisation. The true cost of employment may also be ignored. A publications manager who compares the hourly rate of a subcontracted author with the wages paid to an employee is ignoring other vital costs of employment. Apart from the more measurable costs, such as employer's National Insurance, there's the cost of recruitment, providing benefits, providing desk space, equipment and consumables, insurance and many other related expenses. For a true comparison, the publications manager must be aware of the overall cost to the company of employing the individual. Those not used to doing so could well be surprised by the result. Naturally, the cost of the subcontracted author will depend on their own circumstances, skills and overheads. A documentation house that employs their own authors or sub-subcontracts work, are likely to have much higher overheads than an independent, freelance author.

The publications manager should be wary of employing freelance authors on a continuous basis, especially if they are working on-site. If a company provides facilities to a freelance author on an exclusive contract basis over a significant period of time, the Inland Revenue may consider that the contractor is entitled to employee benefits and should be treated in much the same way as permanent or temporary staff. This may result in the organisation being responsible for Employer's National Insurance contributions and the subcontractors own tax liabilities.

Using subcontracted authors may be the most appropriate solution for those organisations that have only infrequent technical publications requirements. In such circumstances, there may be no justification for employing a full-time author and that subcontracting is the most cost-effective and efficient solution.

The following are a number of points that the publications manager may wish to consider when employing the services of third parties for technical publications:

(a) When choosing a contract writer or documentation house, the publications manager should ensure that their experience is relevant to the particular discipline or subject matter.

(b) It may be helpful to review examples of previous work done, though this can be misleading. A contract author may have to work on jobs with very different quality and content level in order to satisfy the requirements of a particular client. A job done for one client may not reflect the capability of the author; simply the requirement of that particular client. It may not be possible to insist on reviewing other work done by a freelance author or documentation house, particularly if that work is confidential or done for a competitive organisation.

(c) One should establish whether the contractor needs to work on-site. In some cases this is essential if the facilities, contacts and equipment are located on the premises and cannot be provided to a subcontracted author. Working on-site should not be insisted upon as a matter of principle, for example, to 'keep and eye on the author'. This may work to one's disadvantage since the author may be more productive working in their own environment, using their own facilities. They are also not restricted to the working hours of the client.

(d) It should be made clear exactly what the subcontractor is to provide and what the client is to provide. If the author requires access to personnel, equipment or other facilities, these should be written in the contract. If equipment or facilities are to be provided off-site, matters such as insurance and other liabilities should be discussed and agreed before work begins. Confidentiality should also be considered and it may be important for the organisation to obtain written reassurance that all dealings remain confidential, especially if the subcontracting author does similar work in competitive organisations. It is not uncommon to issue confidentiality agreements to cover this, signed by both parties.

(e) It is important that charging rates, expenses and other costs are clearly defined. If the work is carried out on an hourly basis, estimates for the job need to include information about what work is included. For example, the estimated time and therefore cost of the job may not take into account changes required by or incurred by the client. If the subcontracted author is to work on site, travel and accommodation expenses may also be incurred.

(f) Ensure that there are contingencies in the event of the subcontracting author going sick or being unable to complete the documentation project. This has to be treated in the most practical manner; clearly, it is not effective to have another author acting as an understudy, but the publications manager needs to be aware of any procedures that can be put into place in the event of the subcontractor failing to complete his or her responsibilities.

(g) Third-party documentors should be given a clear brief covering all aspects of the tasks expected of them. This should include the level of technical content, extent of illustration, proofing and checking required of them, writing style, design and other elements. Where such information is not given, the publications manager must accept responsibility for the material submitted and has no case for complaint if the documentation submitted does not meet expectations. It must also be said that a publication manager may choose not to provide detailed briefs to subcontractors either because of indecision or in an attempt to deliberately introduce a fresh approach to documentation by allowing freedom of design and expression on the part of the subcontractor. The latter highlights another benefit of using subcontractors for technical documentation.

DOCUMENTATION DEVELOPMENT

Part of the duties of the publications manager is to endeavour to develop the technical documentation process and the documentation itself, for which he or she is responsible, in order to bring about improvement. It is unlikely that any technical publication is as good as it can be and there may be limitations on a development programme outside the manager's control, such as budget and/or timescale.

In the past, the introduction of technology into the technical publications environment has not been cheap, but in more recent years, personal computing has reached such heights that the facilities for providing more control over the publication cycle and improving quality are increasing available and affordable. Although the technology itself offers more possibilities for the technical publications departments, publication managers have a responsibility to their job and to industry to ensure that, by utilising this new technology, standards of quality are maintained and improved. This means directing efforts towards training and gaining knowledge about all aspects of the profession, encouraging good writing, design, presentation, typography, etc.

The training and development of authors under the publication manager's control is an important part of the publication department's development. While there is only small amount of training available to those who want to train authors for commerce and industry, publication managers have a certain responsibility towards encouraging quality and professionalism in documentation. This requires a conscious effort. Development towards better documentation must focus on ways in which the difficulties of communicating ideas and information to readers can be overcome. In this respect, many technical publications have yet to succeed in gaining the attention of the user and then providing the information required. The publications manager needs to be aware of and responsive to new developments and competitive documentation.

REVISION MANAGEMENT

Almost all technical documentation, by the nature of the subject matter which it covers, requires updating. Documentation for technically developing products which may include hardware, software, firmware or other systems, may undergo significant alterations or enhancements. Just how to go about maintaining and updating documentation efficiently requires careful consideration.

There are many factors that influence how to update existing documentation. Economic factors will be high on the list of priorities as well as the time available. Ideally, documentation would be completed revised and replaced when changes become necessary, but this is not always a realistic possibility, particularly if there is already a significant level of stock of the current version of a document that would otherwise be scrapped.

Timing of revisions may be the overriding factor that determines the best way of revising publications. If there is sufficient time to handle amendments, a rewrite of an existing edition may be possible to coincide with the date or release or distribution, but this may still be a costly exercise, particularly if the documentation is subject to frequent changes. Documentation needs to be designed for change. The more frequent the changes, the more flexible the documentation system must be. In the case of software documentation, it may be that, if change is particularly regular, an on-line documentation solution is preferable that does not incur print leadtimes, costs or stockholding problems.

Examples of update documentation show that revisions are handling in a wide variety of ways. In some cases, the publisher issues a separate document, an addendum, that covers all the new and revision material relevant to the next issue. Unless such documentation is integrated into the existing documentation by effective cross-referencing and is well structured, the revision material may well confuse the reader, or may be misunderstood. It is particularly annoying for readers to have to read two versions of a particular subject in order to gain the latest information. Another solution is to provide updated pages for a loose-leaf system of documentation. This may require the reader to remove old pages and replace or insert new ones. This may be the only practical solution for very large documents with few revisions, but it does, nonetheless, cause the user of the documentation more effort. Updating of technical documentation needs to be done as much for the convenience of the readers as it is of the publisher.

There are still many technical publications presented in ring-binders. This offers a major benefit when it comes to revisions as it is possible for either the publisher or the user of the documentation to make changes to pages without affecting other parts of the documentation, particularly if the page numbering system used offers the possibility of amendment or expansion without the need to repaginate the entire publication. Perfect bound or other fixed-bound documentation rules out such a course of action and so the revision must be handled either by providing additional documentation or replacing the current edition in its entirety.

Revisions should be carefully monitored and referenced. If the revision relates to the release of a revised version of a system or product, the documentation revision may be referenced in the same way. For example, software documentation may indicate the current version number of the software to which it relates, or a hardware product to its model number. An amendment history should be maintained and, if page updates or replacements are issued to revise existing documentation, they should be dated and referenced where possible. This not only helps in the collation and distribution of technical publications, but helps when additional support is required as the user of the documentation can quickly identify the level of documentation being used at a given point in time.

Where changes are being made to a technical publication, it can be helpful to provide a summary of the changes made. This is for the benefit of existing users or readers who have already been issued with earlier editions and for whom not all revisions are relevant. For example, if a user manual is revised by supplying a wad of replacement pages, the user may wish to read a summary of the revisions rather than having to read through each page replacement and compare with the original in order to detect the changes or additions supplied. This is more difficult in the case of on-line or multimedia documentation, but even this may be accompanied by a listing of revisions made.

It is important for the author and publications manager to keep a track of revisions made to technical publications. Documents that have been edited and that are to be re-proofed should be clearly identified with a revision date, reference and possibly the source or initials of the person responsible. Some word processing and desktop publishing systems offer a facility called 'redlining' which enables revised text to be printed with a lined or shaded background, or perhaps in a different colour, so that an editor or reviewer of the documentation can detect the revisions easily for checking. Otherwise the editor or reviewer may have to read pages of information that has not been revised in order to locate and check the validity of any new or amended material.

Careful management of text, documentation files and proofs will help to ensure that revisions are handled efficiently. It may also be important to retain such revision information in order to maintain a history in case reference needs to be made to previous levels of updating or additions.

SUMMARY

1. The publications manager has responsibilities much the same as any other departmental manager. He or she must run the publications department in an efficient and effective way to the benefit of the staff, the organisation and the customer.

2. Responsibilities include the recruitment and development of technical publications staff, the control of publications projects, planning, scheduling, costing, evaluating, print buying, quality control and development.

3. The recruitment of authors and other publications staff is one of the key responsibilities of the publications manager. Staff should be trained and developed to their fullest potential.

4. Many publications managers are responsible for selecting printers for technical publications. The print buying activity is one that requires experience as well as education since the selection of the best printer for a particular job can be based on quality and delivery standards as much as on the suitability and cost of facilities offered. Drawing up print specifications for technical publication jobs will help the publications manager in tendering for estimates and retaining a history of printing specifications and prices.

5. As technology develops and more publishing technology becomes available, the options available to the publications manager for in-house publishing widens greatly. Apart from managing systems already in use within a publications department, the manager must be aware of and may be responsible for the procurement of hardware and software systems that improve the effectiveness and efficiency of the department.

6. Costing and scheduling are an important part of the publications manager's role. Costing must be realistic and this often means including indirect costs associated with the publications department as well as simply the salaries of authors, word processor operators, etc. Maintaining budgets for documentation jobs will help the publications manager to control costs and ensure that the most effective solution for a particular job can be afforded with the available budget. Scheduling involves breaking down the various tasks of the individual involved in the publications cycle and planning the time available and used. The schedule for a publications project is often closely related to the development schedule of the project or system for which the documentation is intended.

7. The publications manager may need to consider the subcontracting of writing and other publications services. There are a number of benefits to this, especially if workloads increase suddenly beyond the capacity of the in-house staff and new staff

cannot be trained in sufficient time or is not justifiable financially in the long term. Subcontracted authors can bring specialist skills to the organisation and may be more experience is particular lines of work than in-house staff. They may also be more productive.

8. The development of documentation — including improvements in the publication process as well as the content of documentation — is also the responsibility of the publications manager. This may include introducing new methods and technology, better training for technical publications staff and careful quality control and evaluation of documentation, including any similar or competitive documentation published by other organisations.

10 International Documentation

Documentation that is intended for international audiences requires a different approach to that for a local market. This applies not only to documentation that is to be published in English in other countries, but also that is to be translated into one or more other languages.

WRITING FOR INTERNATIONAL AUDIENCES

One of the most important considerations when writing material intended for international audiences is the choice of phrasing and vocabulary. Use of slang an colloquialisms may result in misunderstanding and misinterpretation. For example, some nouns and verbs are ambiguous and only the precise context in which they are used enables the reader to understand them. The word 'place' is both a noun and a verb and may be misconstrued by those for whom English is a second language. In the following examples, the word has different meanings but they may all be translated as the being the same word:

(a) *Place* the cursor in the top left-hand corner of the screen.

(b) Ensure that the text remains in the same *place*.

(c) Find the *place* in the text where the index reference was entered.

In the first example, *place* is the verb and has the same meaning as put, position, locate, rest, etc., while in the second example, the word *place* has the same meaning as location, position, point, orientation, etc. Already there is a dilemma in which the choice of the most appropriate word is not clear, especially when trying to find the one that translates best in another language. This also highlights another potential problem: consistency. It is all too easy to change the word used to mean a particular thing or action so that the text is not boring to read, but when writing documentation that is to be translated, it is particularly important to use the same words for the same meanings consistently. For example, it is not good practise to use the phrase 'enter the information in this field' in one part of the text and 'input the data in this field' in another part if they are both supposed to mean the same thing. Technological terminology is also important to use consistently. The terms 'VDU', 'CRT', 'Monitor', 'Display', 'Video Display' and 'Screen' can all mean the same piece of equipment. One term should be chosen and used throughout the documentation.

The author should also be aware of the differences between American and British English. Different terms may be used to mean the same thing. A 'lodgement' in American English may be the same as a 'deposit' in financial terms. In electronics, 'Earth' in British English may have the same meaning as 'Ground' in American English. Some words may seem archaic to American readers. For example, the phrase 'whilst holding down the <Ctrl> key...' would be written as 'while holding down the <Ctrl> key...', since the word 'whilst' would be considered almost Dickensian to American audiences.

The following points summarise the things that authors should consider if their documentation is intended for international audiences and may be translated into other languages:

— avoid the use of slang or colloquial phrases;

— be consistent in the choice of terminology and phrasing;

— when in doubt about which term or phrase to use, find out what the translated word or words could mean and ensure the most appropriate form is chosen;

— avoid the use of abbreviations and mnemonics;

— take care in the use of capital letters on words as in some languages, such as German, these may be assumed to be nouns;

— use the international system for units of measurement;

— avoid including anecdotal or humorous content in the text;

— take care with grammatical structures, particularly where text is precied, for example, when annotating illustrations, where verbs or subjects may be omitted.

Apart from getting the wording right for international documentation, the author should realise that, when translated, the documentation may be completely different in extent. That is to say, some languages use a great many more words or longer words to say the same thing as is written in English. Similarly, some languages may be far more economical in the use of page space. This can have implications not only on the cost of the production of the translated documentation, but also on the layout of the page and overall design. A carefully paginated document that includes illustration material which is sent for translation may not be returned in quite the same way and the result may be that considerably more effort is required to prepare artwork suitable for printing.

TRANSLATION

Given that documentation intended for translation must be carefully worded, as described previously, documentation for translation that has not already been prepared with this in mind may require editing before it is passed to a translation bureau.

Apart from editing for consistency and grammatical structure, it may be useful if the author or editor compiles a list of specialist technical terminology that has been used within the documentation. This can be given special attention by the translators to ensure that the correct translation is used in the appropriate context.

There are many translation agents and bureaux to choose from, but care should be taken to ensure that they are familiar with the area of specialisation in which the documentation is intended for use. Most agents identify their particular areas of technical specialisation (such as medical, electronics, mechanics, avionics, mining, etc.) but it may be worth evaluating examples of previously translated material, if available.

Generally speaking, translation should not be done from English into a foreign language by individuals for whom that foreign language is not native. It is inevitable that, as a consequence, the translation will read as though it has been translated badly, however grammatically accurate, since the phraseology will probably seem stilted and unnatural. Some organisations will only use translators in the country in which the translated documentation will ultimately be used. Others may prefer to have the text translated by a local agency, but vetted and edited by a native of the destination country so that irregularities can be ironed out.

One of the more difficult aspects of translated technical documentation involves its checking and proofing. Clearly, a document that has been handed to a translation bureau for will be returned in a manner which then makes it unreadable to the client who cannot, at first hand, check that the translation is accurate and that the correct meaning of the content has been retained. This may result in the need to engage the services of an independent checker who not only knows the language, but also understands the content to the appropriate technical level. The document can either be checked and approved or edited by that individual or be reverse-translated back into the original language. Of course, the content can be misinterpreted during either operation and there is no necessarily a simple and economical solution to this.

SUMMARY

1. Special care needs to be taken when writing or preparing technical documentation for international audiences, particularly if it is intended for translation into other languages. The author should be consistent in the use of terminology and grammatical structure and avoid ambiguous, colloquial and slang expressions or terms.

2. The translation of technical documentation must be carried out by qualified individuals, preferably whose native tongue is that of the language into which the document is to be translated and who has sufficient technical knowledge to be able to understand the content.

FURTHER READING

Austin, M. *Technical Writing and Publication Techniques* (ISTC, 1987).

11 Training and Education

AUTHOR TRAINING

There is little academic training available for technical writers that specialises in the subject of authorship. This makes the selection of author candidates for recruitment difficult to base on a recognised qualification. There are a few courses that result in a certificate or diploma, a City and Guilds course and some degree courses that include topics relevant to technical authors under the general term of 'Technical Communications'. There are also a variety of seminars and short courses that are useful for both potential and existing technical authors. All courses are useful not only for those intending to take up technical authorship as a career, but also for those with a technical background who are already involved in documentation or technical writing and who have not had any formal tuition in this field so far.

NATIONAL QUALIFICATIONS

The following describes some of the nationally recognised education courses relating to technical documentation personnel.

City and Guilds

The main qualifications for technical authors and publications managers are covered by the City and Guilds 536 certificates. These are divided into two parts: scheme 536/1 covering publication techniques and scheme 536/2 dealing with technical authorship which includes optional specialisation in mechanical and electrical disciplines. Typically, part 1 is taken before part 2, but this is not obligatory. Training for the City and Guilds certificate is available from various sources including full- or part-time courses at colleges and universities, or by distance learning (correspondence courses), some of which are identified in this chapter.

Courses leading to the City and Guilds examinations but covering other subject matter are also available, as well as courses of equivalent or more advanced level for particular disciplines not covered by the City and Guilds options. Some of these courses can lead to college or other certificates and tuition is available full- or part-time or by distance learning.

For details of application for the City and Guilds examinations, write to:

City and Guilds of London Institute
46 Britannia Street
London
WC1X 9RG

Higher Education

BTEC Higher National Certificates and Diplomas, degrees and post-graduate studies are available in this sector. Tuition is normally full-time, sometimes with day-releases for study while working; part-time or block-release study is possible at post-graduate level. Education and training for technical illustration is predominantly available through these courses which are normally entered from school and taken full-time.

DISTANCE (OPEN) LEARNING

For those individuals already working in commercial or industrial environments, full- or part-time training may be difficult unless the organisations for whom they work are prepared to give leave to attend. By using distance learning (correspondence courses), that follow the principles of the Open University, students can follow a flexible system with learning wherever convenient, whether at home or in the work place. Costs for distance-learning courses can, in some instances, be cheaper too, and as there are no accommodation expenses involved, more students can be trained in this way for a given budget.

With distance-learning courses, the emphasis is on learning rather than teaching and the student may work at his or her own pace, so can spend more time on aspects that he or she finds slower to assimilate, without being left behind as occurs in the classroom. The student also benefits from a one-to-one attention of the tutor.

INDEPENDENT COURSES

These include short courses and seminars designed to help train and retrain potential or existing technical authors respectively, in various aspects of technical documentation production and communication skills. Many of them take the form of one or more day's full-time attendance and offer an ideal opportunity to meet with industry peers.

COURSES IN TECHNICAL COMMUNICATION

There are a number of full-, part-time and distance-learning courses relating to technical communications available in colleges, universities and through independent organisations around Britain. The following lists offer a guide to some of these courses. It is important to note that this information can vary each year and you are advised to check with the organisation concerned about availability of any courses mentioned here or others they may have in addition or as replacements.

The list is not exhaustive. Apart from contacting your local colleges, you may be able to obtain more up-to-date information on relevant courses from the Institute of Scientific and Technical Communicators at the following address:

ISTC
Kings Court
2/16 Goodge Street
London W1P 1FF

Telephone: 0171 436 4425

148   TECHNICAL DOCUMENTATION

Technical Communication — Full Time

Title	Organisation	Qualification	Duration
Communication Studies	Dundee College of FE	Scotvec	1 year
Technical Authorship	Intereurope Technical Services Ltd	CGLI 536-1, CGLI 536-2 & Diploma	8 weeks
Technical Authorship	Grantham College	CGLI 536-2	1 year
Technical Authorship	Chippenham Technical College	CGLI 536-1, CGLI 536-2 & Certificate	1 year
Technical Authorship	Chequers Bureau Technical Services	CGLI 536-1, CGLI 536-2 & Diploma	8 weeks
Engineering Tech. Auth.	Chippenham Technical College	CGLI 536-1, CGLI 536-2	1 year
Technical Comm.	Coventry Polytechnic	Honours Degree	3 years
Technical Comm.	Blackpool & Fylde College	BTEC HNC	33 weeks
Tech. Comm. Techniques	Grantham College	CGLI 536-1	variable
Technical Writing	Blackpool & Fylde College	BTEC HND	3 years

Technical Communication — Part Time

Title	Organisation	Qualification	Duration
Communication Studies	Newcastle upon Tyne Polytechnic	Post-graduate Diploma	3 × 1 mnth
Communication Studies	Sunderland Polytechnic	MA/Post-graduate Diploma	variable
Communications 1, 2, 3	Clydebank College	Scotvec	40 hours
Communications 4	Clydebank College	Scotvec	80 hours
Engin. Comms.	Clydebank College	Scotvec	40 hours
Technical Authorship	Chippenham Technical College	CGLI 536-2	36 weeks
Technical Authorship	Dundee College of FE	CGLI 536-2	variable
Technical Authorship	Highbury College of Technology	CGLI 536-2	180 hours
Tech. Comm. Techniques	Highbury College of Technology	CGLI 536-1	36 weeks
Tech. Comm. Techniques	Grantham College	CGLI 536-1	variable

Technical Communication — Distance Learning

Title	Organisation	Qualification	Duration
Technical Authorship	Intereurope Technical Services	CGLI 536-1, CGLI 536-2	variable
Technical Authorship	Tutortex Services	CGLI 536-1, CGLI 536-2	variable
Technical Authorship	College of Technical Authorship	CGLI 536-1	150 hours
Technical Authorship	College of Technical Authorship	CGLI 536-2	100 hours

Miscellaneous — Full Time

Title	Organisation	Qualification	Duration
Access to IT	Chequers Bureau Tech Services	RSA Stage 1	4 × ½ days
Basic Tech. Authorship	Pergamon TSIB BV		2 weeks
Advanced Tech. Author.	Pergamon TSIB BV		1 week
Basic Tech. Writing	Chequers Bureau Tech Services		7 days
Intro to DTP	Chequers Bureau Tech Services		1 day
Intro to Tech. Writing	Pergamon TSIB BV		3 days
Report Writing & Edit.	Chequers Bureau Tech Services	Bureau Certificate	4 days

TRAINING AND EDUCATION

Miscellaneous — Full Time

Title	Organisation	Qualification	Duration
Tech. Auth. Short Course	Chequers Bureau Tech Services		variable
Tech. Auth. Short Course	Intereurope Technical Services		2 weeks
Tech. Pubs Management	Pergamon TSIB BV		4 days
Various Technical Documentation courses	The John Kirkman Communication Consultancy		various
Report Writing for Engineers	Chippenham Technical College		various
Teaching Media Studies	South Nottinghamshire College of FE		1 year
Various Technical	Informatics Resource Centre		various

Miscellaneous — Part Time

Title	Organisation	Qualification	Duration
Technical Writing	City University		15 weeks
Technical Writing	Extension Studies Unit, City Univ.		12 weeks
Writing in Industry and Commerce	City University		10 meetings
Writing Publicity Copy for Technical Authors	Chippenham Technical College	Intermediate	various

Miscellaneous — Distance Learning

Title	Organisation	Qualification	Duration
Microcomputer Technology	Tutortex Services	Tutortex Certificate	variable
Report Writing	Telford College	Basic	2×3 hours

Design — Part Time

Title	Organisation	Qualification	Duration
Graphic Communication	Nene College	CGLI 524	2 years
Graphic Design	Cleveland College of A & D	BTEC National Certificate	2 years
Graphic Design	Farnborough College of Technology	BTEC National Certificate	3 years
Graphic Design	Farnborough College of Technology	BTEC HNC	2 years
Graphic Design	Manchester Polytechnic	BTEC HNC	2 years
Graphic Design	Nene College	BTEC HNC	2 years
Graphic Design	South Notts College of FE	BTEC National Diploma	2 years
Graphic Design	West Nottinghamshire College	BTEC National Diploma	3 years +
Industrial Design	West Nottinghamshire College	BTEC National Certificate	3 years +
Technical Illustration	Bath College of FE	BTEC National Certificate	2 years
Technical Illustration	Birmingham College of A & D	BTEC National Certificate	2 years
Technical Illustration	Birmingham College of A & D	BTEC HNC	2 years
Technical Illustration	Richmond upon Thames College	BTEC National Certificate	3 years
Typography	South Fields College of FE	BTEC HNC	1–3 years

150   TECHNICAL DOCUMENTATION

Scientific & Technical Illustration — Full Time

Title	Organisation	Qualification	Duration
Graphic Information Design	Falmouth School of A & D	CNAA BA Honours	4 years
Scientific Illustration & Image Processing	Southampton Institute of HE	Institute Diploma (SIAD)	2 years
Technical Illustration	Bournem'th /Poole College of A & D	BTEC National Diploma	2 years
Technical Illustration	Doncaster Metropolitan Inst. of HE	BTEC National Diploma	2 years
Technical Illustration	Falmouth School of A & D	BTEC National Diploma	2 years
Technical Illustration	Isle College	BTEC National Diploma	2 years
Technical Illustration	Mid-Cheshire College of FE	BTEC National Diploma	2 years
Technical Illustration	Mid-Warwickshire College of FE	BTEC National Diploma	2 years
Technical & Information Illustration	Bournem'th /Poole College of A & D	BTEC HND	2 years

Some College Addresses

The following are addresses of some of the colleges and organisations including those mentioned in the above listings.

Bath College of Further Education, Avon Street, Bath, BA1 1UP

Birmingham College of Art and Design, Linden Road, Bournville, Birmingham, B30 1JX

Blackpool and Fylde College, Faculty of Arts, Dept of Visual Communication, Palatine Road, Blackpool, FY1 4DW

Bournemouth and Poole College of Art and Design, Wallisdown Road, Poole, Dorset, BH12 5HH

Chequers Bureau Technical Services Ltd, 32a Chequers Court, Huntingdon, Cambs, PE18 6LP

Chippenham Technical College, Cocklebury Road, Chippenham, Wilts, SN15 3QD

City University, Extension Studies Unit, Northampton Square, London, EC1 0HB

Clydebank College, Kilbowie Road, Clydebank, G81 2AA

College of Technical Authorship, PO Box 7, Cheadle, Cheshire SK8 3BY

Coventry Polytechnic, Priory Street, Coventry, CV1 5FB

Doncaster Metropolitan Institute of Higher Education, Dept of Art and Design, Church View, Doncaster, S Yorks, DN1 1RF

TRAINING AND EDUCATION 151

Dundee College of Further Education, Melrose Terrace, Dundee, DD3 7QX

Grantham College, Stonebridge Road, Grantham, Lincs, NG31 9AP

Highbury College of Technology, Dover Court Road, Cosham, Portsmouth, Hants, PO6 2SA

Intereurope Technical Services Limited, High Walls, East Street, Fareham, Hants, PO6 2SA

Isle College, Ramnoth, Road, Wisbech, Cambs, PE13 2JE

The John Kirkman Communication Consultancy, PO Box 106, Marlborough, Wiltshire, SN8 2RU

Centre for Applied English Studies, University of Wales, PO Box 94, Cardiff, CF1 3XE

Falmouth School of Art and Design, Woodlane, Falmouth, Cornwall, TR11 4RA

The Informatics Resource Centre, 2 The Chapel, Royal Victoria Patriotic Building, Fitzhugh Grove, London, SW18 3SX

Mid-Cheshire College of Further Education, Hartford Campus, Northwich, Cheshire, CW8 1LJ

Mid-Warwickshire College of Further Education, Warwick New Road, Leamington Spa, Warwickshire, CV32 5JE

Newcastle upon Tyne Polytechnic, Communications Unit, Faculty of Arts & Design, Newcastle upon Tyne, NE1 8ST

Nene College, Avenue Campus, St Georges Avenue, Northampton, NN2 6JD

Pergamon TSIB BV, Zwaluwlaan 1, 2211 LD Noordwijkerhout, The Netherlands

Richmond upon Thames College, Egerton Road, Twickenham, Middlesex, TW2 7SJ

Southampton Institute of Higher Education, East Park Terrace, Southampton, SO9 4WW

South Fields College of Further Education, Aylestone Road, Leicester, LE2 7LW

South Nottinghamshire College of Further Education, Farnborough Road, Clifton, Nottingham, NG11 8LU

Sunderland Polytechnic, Edinburgh Building, Chester Road, Sunderland, SR1 3SD

Telford College, Crewe Toll, Edinburgh, EH4 2NZ

152 TECHNICAL DOCUMENTATION

Tutortex Services, 55 Lightburn Avenue, Ulverston, LA12 0DL

West Nottinghamshire College, Derby Road, Mansfield, Notts, NG18 5BH

Details extracted from information supplied by the Institute of Scientific and Technical Communicators. Contact your local college or university for information about courses offered relevant to technical documentation. See also the list of organisations and technical publication services later in this book, some of whom also offer short, independently run courses.

12 List of Standards

The British Standards Institution (BSI) is an organisation incorporated by Royal Charter and represents the U.K.'s voice in the European standards area. BSI subscribing membership provides access to all standards information and helps companies develop products and services in accordance with British and European standards. Members are entitled to *BSI News*; a monthly journal which announces all new and revised standards produced throughout the year. A member's catalogue is also published and a discount of 50% is available on purchases of standards. Applications can be made to the following address:

BSI
Linford Wood
Milton Keynes MK14 6LE
Telephone: 01908 220022

STANDARDS

The following lists some of the British Standards published by the British Standards Institute (BSI) and are relevant to technical documentation. There are others that cover details in specific disciplines, but the following are generally applicable. Although not particularly cheap, especially to non-subscribing members, they should become part of the technical author's library. It is important to note that standards are updated. For example, BS4884: 1992 Part 1 Technical Manuals, and 1993 Parts 2 and 3 Technical Manuals supersedes the former 1973 editions and highlights the need to check the latest publication date of any standards documentation already in one's possession. New standards are also emerging. One of the most recent being the BS7649: 1993, Guide to the design and preparation of documentation for users of application software, which covers an area of technical documentation that has, for many years, had no specific standards to follow.

BS308	Engineering drawing practice
BS350	Conversion factors and tables
BS530	Graphical symbols for telecommunications
BS974	Symbols for use on flow diagrams of chemical and petroleum plant
BS1219	Preparation and correction of mathematical copy and proofs
BS1311	Manufacturers' trade and technical literature (sizes)
BS1413	Page sizes for books
BS1523	Glossary of terms used in automatic controlling and regulating systems
BS1553	Graphic symbols for general engineering
BS1629	Bibliographical references
BS1635	Graphical symbols and abbreviations for fire protection drawings
BS1646	Graphical symbols for process measurement and control functions and instrumentation
BS1896	Sizes of reprographic papers
BS1991	Letter symbols, signs and abbreviations
BS2517	Definitions for use in mechanical engineering
BS2856	Precise conversion of inch and metric sizes on engineering drawings
BS2917	Graphical symbols for use in diagrams for hydraulic and pneumatic systems
BS2961	Typeface nomenclature and classification
BS3203	Glossary of paper, board, pulp and allied terms
BS3238	Graphical symbols for components of servo-mechanisms
BS3363	Schedule of letter symbols for semi-conductor devices and integrated circuits
BS3527	Glossary of terms relating to automatic data processing
BS3641	Symbols and machine tools
BS3700	Preparation of indexes
BS3763	International System (SI) units
BS3811	Glossary of maintenance terms in terotechnology

LIST OF STANDARDS

BS3939	Graphical symbols for electrical power, telecommunications and electronics diagrams
BS4000	Sizes of papers and boards
BS4884	Technical Manuals
BS4899	Guide to user's requirements for technical manuals
BS5097	Specification for loose-leaf binders; Part 1: Ring Binders with metal mechanisms
BS5444	Recommendations for preparation of technical drawings for microfilming
BS5605	Recommendations for citing and referencing published material
BS5641	Recommendations for loose-leaf publications
BS5760	Reliability of systems, equipment and components
BS5775	Specification for quantities, units and symbols
BS6046	Use of network techniques in project management; Part 1: Guide to the use of management, planning review and reporting procedures
BS7581	Recommendations for presentation of tables and graphs
BS7649	Guide to the design and preparation of documentation for users of application software

13 Organisations and Institutions

There are many organisations and institutions which may have relevance in the field of publishing and technical writing. Some key organisations are included in this chapter.

THE INSTITUTE OF SCIENTIFIC AND TECHNICAL COMMUNICATORS

The Institute aims to establish and maintain professional codes of practice for people engaged in all branches of scientific and technical communication. It provides a forum for the exchange of views between its members and aims to further their expectations and interests.

The membership embodies a wide range of specialised knowledge of the principles and modern practices of scientific and technical communication. Through its publications and its meetings, the Institute disseminates this experience to a growing profession and those who employ the services of its members.

History

The first organisation for people engaged in technical communication was the Presentation of Technical Information Group, formed in 1948. The Technical Publications Association came into being in 1953 and later changed its name to the Institute of Technical Authors and Illustrators. A further organisation — the Institute of Technical Publicity and Publications — was formed in 1963. In 1972, these three organisations merged to form the Institute of Scientific and Technical Communicators, thus combining the formality of a professional institute with the informality of a technical society.

Members

The Institute includes amongst its members specialists whose activities are concerned with the communication of technical information. Although the Institute was in the past based on industrial technical publications, today about half its members are engaged in that area.

ISTC members include a wide range of other technical communication specialists and people who are concerned with the selection and use of various communication media.

Members are employed in industry, commerce and finance, government and educational establishments, the armed services and specialised contracting organisations and consultancies.

Administration

The Institute is governed by a Council, which also appoints the Officers. Implementing the Council's policy is the responsibility of the Board of Management, which comprises the Institute's Officers and the chairmen of its committees (each of whom is a member of the Council).

The Institute's committees are responsible for the following areas of interest:

— education and training

— marketing and external relations

— programmes and membership services

— membership recruitment and admissions

— publications and technical information

— professional practices and discipline.

Membership Grades

Anyone who shares the aims and interest of the Institute is welcome to join. There are five grades of membership.

Corporate members are either Fellows or Members and they are entitled to use the designations FISTC or MISTC following their names. Non-corporate members are Associate Members, Students and Companions.

The grade of membership is based on age, experience and responsibility in the field of technical communication. As a guide, Fellows must be at least 35 years old, have advanced qualifications or experience and be employed in positions of responsibility. Members must be at least 26 years old and have significant qualifications, experience and responsibility.

Associate members are people who do not yet have experience for a corporate grade. Students are people undergoing further or higher education and training. Companions are people who have a close interest in scientific and technical communication, but who are not employed wholly as communicators. Full details of qualifications required for the grades of membership can be obtained from the General Secretary at the address below.

Publications

The Communicator is the journal of the Institute. It contains articles of professional interest, as well as news of the Institute's activities and of members. It is published four times a year, free of charge to members; it is also available by subscription. The Institute's handbook of technical writing and publication techniques, first published in 1985, was updated in a second edition in 1990.

Meetings

Evening meetings are held in London between September and May. Visits, conferences and seminars are arranged as opportunities arise, and the Institute arranges an annual lecture given by a noted communicator. Out-of-London meetings are assuming increasing importance. The Institute also collaborates with other bodies in mutually beneficial activities.

Education and Training

The Institute presents the views of its membership to government departments dealing with education, and it liaises with universities, polytechnics, colleges and such national bodies as the National Council for Vocational Qualifications and the City and Guilds of London Institute, to advise on education and training. The Institute recognises courses and examinations in relation to membership and offers awards for people whose work and examination results are judged exceptional.

Intecom

The ISTC is the United Kingdom's member of INTECOM (International Council for Technical Communication), an international association of technical communication societies. INTECOM's main activity is the promotion of an international forum for the exchange of views on technical communication, held in various countries, and organised by one of the member societies.

> The Institute of Scientific and Technical Communicators
> Kings Court 2/16 Goodge Street
> London
> W1P 1FF
> Telephone 0171 436 4425
> Fax 0171 580 0747

THE SOCIETY OF AUTHORS AND THE SCIENTIFIC AND TECHNICAL AUTHORS GROUP

The Society of Authors offers a variety of benefits and services to professional authors. Although primarily aimed at authors in the general field of publishing, the Society also has specialist groups, which include the Science, Technical and Specialist Group who publish their own bulletin. This specialist group within the Society of Authors reports its activities in the Noticeboard section of *The Author* — the journal of the society — and publish their own official newsletter called *The Stag*.

The Society campaigns for the benefit of authors in many ways including negotiating improved terms and conditions; lobbying MPs, ministers and government departments for new legislation (for example, Copyright Act); promoting the interests of specialist writers; supporting the British Copyright Council; Backing the Author's Licensing and Collecting Society; litigating in matters of importance to authors and organising regional groups.

The Society was founded by Walter Besant in 1884 with Lord Tennyson as the first president, to prompt the interest of authors and to defend their rights. A great many prominent writers, including Sahe, Galsworthy, Hardy, Wells, Barrie, Masefield, Forster, A. P. Herbert and countless contemporary writers, have assisted in the activities and campaigns of the Society.

The Management Committee consist of 12 professional authors, who are elected and serve for three years. The Council is an advisory body which meets twice a year. There are over 5500 members, representatives of all the media.

The permanent staff has great experience of the business aspects of professional authorship. In addition, the Society has immediate access to solicitors, accountants and insurance consultants.

The Society of Authors is a limited company and has been certified as an independent trade union (not affiliated to the TUC).

Although the Society is financed primarily by subscriptions, an important part of its income comes from handling the literary and dramatic rights for the estates of such eminent authors as Bernard Shaw, T. S. Eliot and E. M. Forster.

The Society administers a number of prizes and grants. Through their permanent staff, the Society is able to offer personal service, helping writers in the following ways:

— advising on negotiations;

— taking up complaints;

— pursuing legal actions;

— supplying a quarterly journal *The Author* and a twice yearly supplement, *The Electronic Author*;

- publishing and supplying guides to members free of charge covering a wide range of topics including copyright and moral rights, income tax, publishing contracts, advice on purchasing a word processor, etc.;
- inviting members to conferences, meetings and social occasions.

Members also have access to the following facilities:

- book ordering scheme to enable members to purchase books from most major publishers at a trade discount through the Society;
- retirement benefit scheme;
- the Author's Licensing and Collecting Society to which members are automatically entitled to free membership;
- group medical insurance schemes with both BUPA and the Bristol Contributory Welfare Association;
- a contingency fund which provides financial relief for authors and their dependants in sudden financial difficulties;
- a pension fund which offers pensions to a number of members over 60 who have belonged to the Society for at least ten years;
- the Royal Over-Seas League offering concessionary subscriptions rates to Society members;
- photocopier facilities;
- membership cards entitling members to Reader's Tickets for the British Library.

Further details can be obtained from:

The Society of Authors
84 Drayton Gardens
London SW10 9SB

SOCIETY OF FREELANCE EDITORS AND PROOFREADERS

The Society was founded in 1988 in response to the growing number of freelance editors and their increasing importance to the publishing industry. The Society aims to promote high editorial standards by disseminating information and through advice and training and to achieve recognition of the professional status of its members. The Society also supports moves towards recognised standards of training and qualifications.

Further details can be obtained from:

The Secretary
Society of Freelance Editors and Proofreaders
16 Brenthouse Road
London E9 6QG

SOCIETY OF INDEXERS

The Society, founded in 1957, publishes *The Indexer* biannually and *Micro-Indexer* another biannual dealing with computer indexing and a quarterly newsletter. The Society issues an annual list of members and the (*IA*) *Indexers Available*, which lists members and their subject expertise. In addition, the Society runs an open-learning course entitled Training in Indexing and recommends rates of pay.

Further details can be obtained from:

Society of Indexers
38 Rochester Road
London NW1 9JJ

TECHNICAL DOCUMENTATION SERVICES

The following is a list of organisations specialising in technical documentation and publishing services. This list is by no means exhaustive. Other services may be found advertising in various journals and magazines and may be included in *Yellow Pages*. None of the services listed here are recommended in any way by either the author or NCC Blackwell and no responsibility is accepted for their quality of service. They are provided purely for information purposes only.

ACD Documentation and Training

Elsden House
47 Elsden Road
Wellingborough
Northants
NN8 1QD

Telephone 01922 223311

Freelance technical authorship of training and user-guides for commercial computer applications.

Adept Writing Services Limited

Bulldog House
London Road
Twyford
Reading
Berkshire
RG10 9EU

Telephone 01734 344800

ORGANISATIONS AND INSTITUTIONS 163

Offering a complete documentation service for computer systems and company procedures, context-sensitive help systems and expertise for a range of Windows and stand-alone hypertext tools. Also able to convert paper-based documents into hypertext and provide consultancy, authoring and individual training.

BDC Technical Services

Slack Lane
Derby
DE22 3FL

Telephone 01332 347123

Offering a wide range of services including technical documentation, database management, storage and retrieval software, publishing services and full print facilities. Specialising in procedural manuals and electronic delivery documentation. World leaders in on-line documentation in many formats such as CD-ROM.

Brooklands Design Services Limited

Park House
Greenhill Crescent
Watford
Herts
WD1 8QU

Telephone 01923 229234

Fax 01923 299234

Placement of permanent contract technical authors with skills in hardware, software and electromechanical documentation. Applications covered are communications, military and industrial. Approved to ISO 9002.

Corby Communications Limited

PO Box 2395
Birmingham
B12 8LL

Telephone 0121 446 4711

Fax 0121 446 4712

Services offered include recruitment, permanent and contract, software, comms, telecoms, illustrators, authors. Documentation for user-guides, specifications, marketing and sales literature. Training in Ventura, PageMaker, Freelance, Illustrator, Corel Draw, Freehand, Designer.

CPC Lithographic Printers

1 Trafalgar Place
Fratton
Portsmouth
Hampshire
PO1 5JJ

Telephone 01705 752621

Fax 01705 831056

Specialist litho printing and bulk photocopying operating country-wide in the field of short-run technical handbooks and manuals, software documentation and training material. Comprehensive facilities from origination through to extensive finishing and binding. The service includes the supply of all peripherals, ring binders, tabbed dividers and most types of software packaging.

CTP Logistics Limited

Montpellier House
Suffolk Place
Cheltenham
GL50 2QG

Telephone 01242 519119

Electronic network technical documentation. Authoring, illustrating, parts catalogues, codification, microfiche, printing, ILS (integrated logistic support).

Digitext

15 High Street
Thame
Oxon
OX9 2BZ

Telephone 01844 214690

Fax 01844 213434

CompuServe 100414,3205

Digitext provides documentation and training solutions for IT systems suppliers and users throughout Europe, using its own proven methodology. Services include on-line and traditional documentation projects, translation, consultancy and provision of contract and permanent technical authors. Training courses include the widely acclaimed 'How to Write On-line Help for Windows'.

Europa Technical Translations

Lloyds Bank Chambers
116 High Street
Smethick
Warley
West Midlands
B66 1AA

Telephone 0121 565 3030

Fax 0121 565 1935

Modem 0121 555 6054

Translations from and into all languages. Word processing in all languages. Electronic publishing in all languages using in-house system. Typesetting, proof-reading, artwork preparation and printing of English and foreign language publications.

EuroTech Publishing

The Anchorage
Swaffham Road
Ashill
Thetford
Norfolk
IP25 7BT

Telephone 01760 440076

Fax 01760 440076

Efficient and flexible solutions to technical documentation problems. Typical projects include software documentation, operating and maintenance manuals, quality assurance, administration and safety procedures. Clients range from a two-man companies to the European Commission, and include the oil industry, financial institutions and mechanical and electronic equipment manufacturers.

Fircroft Engineering Services Limited

Trinity House
114 Northenden Road
Sale
Cheshire
M33 3HD

Telephone 0161 905 2020

Fax 0161 969 1743

166 TECHNICAL DOCUMENTATION

Specialise in the provision of on-site engineering/technical personnel. Also offer both on-site and in-house technical publications service including technical authors, technical illustrators and word processing. Teams provided short or long term. Permanent staff section dealing with all engineering disciplines as well as accountants.

Holly Oak Print

Unit 10, Bakers Lane
Norton, Daventry
Northants
NN11 5EL

Telephone 01327 78217

Fax 01327 705926

Specialist authors of literature for all medical and allied subjects, including electronic and computer applications in diagnostic and therapeutic medicine (expertise in general medicine, physiology, pharmacology, haematology, biochemistry, electronics, instrumentation and computing). Design, typesetting and printing of all technical literature, translation of technical material to and from most languages.

Howard UK Limited

Mitchell House
Southampton Road
Eastliegh
Hants
SO50 5PA

Telephone 01703 644554

Fax 01703 644502

Technical documentation (software and hardware) training documentation and systems, CAL-compliant documentation, information retrieval systems, CD-ROM applications, QA and safety documentation, proposals preparation, illustration and artwork, media conversions.

Industrial Authoring

251 Greys Road
Henley-on-Thames
Oxon
RG9 1QS

Telephone 01491 579926

Fax 01491 578278

ORGANISATIONS AND INSTITUTIONS   167

Desktop publishing. Authoring disciplines: software/hardware user-guides, electronic/mech. installation and repair guides, illustrated parts catalogues, general industrial procedures, document-to-disk conversion, scanning service (text and illustration) documentation consultancy and editing service.

Illustrated Technical Publications

15 Montrose Way
Thame
Oxon
OX9 3XH

Telephone 0184 421 6186

Hardware and software user and service manuals. Assembly instructions, illustrations, diagrams and graphs, scanning (OCR text and graphics), high-quality laser printing.

JW Services

2 Frederick Place
Wokingham
Berkshire
RG41 2YB

Telephone 01734 788322

Fax 01734 788322

Mobile 0378 332267

Technical author (electronics) user and technical manuals, circuit diagram drafting, OCR and image scanning, desktop publishing.

Kudos

9/10 Westminster Court
Hipley Street
Woking
Surrey
GU22 9LQ

Telephone 01483 747227

Fax 01483 747337 and 01483 771279

Compuserve 71612,1545

Providing a complete documentation service, including information design, bespoke training, translation and localisation, project management, contract staff, recruitment, multimedia development, on-line and hypertext.

168 TECHNICAL DOCUMENTATION

Marlow Durndell

18–20 Chapel Street
Titchmarsh
Kettering
Northants
NN14 3DA

Telephone 01832 734950

Fax 01832 735122

Technical documentors to the software industry. User manuals, help text, tutorials, etc. Authors of almost all business user documentation for The Sage Group plc, Pegasus Software Limited, Systems Union, Exact Software (UK) Limited, Atlantic Coast plc, FCCS Limited, among many others. Specialist documentation for software applications (clients include Mobil Oil Company). Production of documentation from any level, including text only through to colour-separated film-setting. Over 400 user manuals written and produced to-date.

Peterborough Technical Communication

2–4 High Tech Centre
Bakewell Road
Orton Southgate
Peterborough
PE2 6XU

Telephone 01733 237037

Fax 01733 239849

European specialists in multilingual technical documentation, technical authoring, recruitment, software translation, consultancy, marketing communications and training.

Printech International Group Limited

Cloverhill Industrial Estate
Clondalkin
Dublin 22
Ireland

Telephone 00 353 1 4573200

Fax 00 353 1 4573256

Specialist printers to the computer and electronics industry. Full translation and localisation service, fulfilment and kitting, component sourcing and testing, consolidation, inventory control and distribution.

ORGANISATIONS AND INSTITUTIONS   169

Technical Graphics Limited

2 Wyvern Avenue
Greg Street
Reddish
Stockport
SK5 7DD

Telephone 0161 477 6645

Fax 0161 477 7657

Publications support over a wide range of areas such as authorship, illustration, typesetting, photography, workshop and print with strong commitment towards new technology. Around the clock cover of services when necessary. Recognised MOD approval.

TMS Computer Authors Limited

Hambledon House
Catteshall Lane
Godalming
Surrey
GU7 1JJ

Telephone 01483 414145

Fax 01483 419717

Provides documentation, recruitment, training and translation services to users and suppliers of the IT industry. Project service: design, preparation and suppliers of the IT industry. Project documentation. Contract and recruitment services: information specialists on permanent or contract basis. Training services: preparation and presentation of materials and courses. Translation and software localisation services.

Wytech Publications Limited

363–367 Sticker Lane
Bradford
West Yorkshire
BD4 8RJ

Telephone 01274 684818

Fax 01274 651345

Experience covering subjects ranging from computers, modems and software to mining machinery. Technical authorship, illustration, graphic design and artwork for leaflets, brochures, manuals and training aids. Exploded and cutaway illustrations, airbrush and wash drawings. Typesetting and professional DTP.

Glossary

The following are some of the terms associated with desktop publishing and typography that the technical author or publications manager may find useful.

ALIGNMENT

This refers to the position of text on a page or within a column. It may be aligned left (ranged to the left and ragged right), centred, aligned right (ranged right and ragged left) or justified (text lined up on both left and right). See also *Quadding*.

ANCHOR

A specific position on a page may be defined as an anchor point to which, for example, a graphic may be associated. This means that if the layout of the page changes the graphic stays fixed to its specified anchor point.

ASCENDER

This is the part of a lower case letter that rises above its main body or x-height (see also *x-Height*), with such letters as b, h and d. See also *Descender*.

BASELINE

Different typefaces vary in size and style and, so that they can be mixed on the same line, are aligned on an imaginary horizontal reference line known as the baseline. See example shown below.

BIT-MAPPING

This relates to the building up of an image using a matrix of dots. Scanners and laser printers use bit-mapping for processing graphics. The quality of the image produced is dependent on the resolution or density of dots in a given area. This is usually expressed as a number of dots per inch, for example 600 dpi.

BLEED

A graphic image may be placed on a page such that it extends beyond the page edge. If the document is reproduced by a commercial printer, the excess paper is cut off when the page is trimmed to size, and the graphic consequently extends to the very edge of the paper.

BROMIDE

This is the photo-sensitive paper on which the output of a phototypesetter is developed through a processor, not unlike a film processing system. In fact, the term 'film-setting' is still used by some people, and the bromides are referred to as the film output.

CICERO

This is a unit of measurement that is common in Europe. It is equivalent to 4.55 millimetres or 0.178 of an inch and is used for measuring type. See also *Point System*.

CLIPBOARD

This is where text or graphics that have been 'cut' are temporarily stored by DTP systems. See also *Cut and Paste*.

CONDENSED TYPE

This is not often available on DTP systems but is found as a facility on some digital phototypesetters. The term itself refers to the relative narrowness of characters in a particular typeface. For a digital typesetter, changing what is known as the 'set width' can invoke the condensing of the characters. For example, a 14-point character can be defined as having a set width of 12 points, which reduces the size of the character's width rather than simply adjusting the space between the characters.

CROP MARKS

These are the small lines that indicate the corners of a page when the page is smaller than the paper on which it is printed. They are used as guides for trimming the page to its correct size and may also be referred to as tick or trim marks.

CUT AND PASTE

This phrase, used often in DTP systems, describes the facility of 'cutting' a piece of text or graphic, therefore removing it from the page, and then 'pasting' it in another place. Before the advent of DTP and WYSIWYG screens this process used to be done manually, literally cutting and then pasting into position with adhesive.

DESCENDER

This is the part of a lower case character that hangs below its main body or x-height (see also *x-Height*), with such letters as g, y and p. See also *Ascender*.

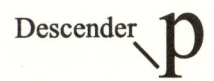

Descender ⟍p

DISCRETIONARY HYPHEN

This is a hyphen, placed in a word at an appropriate point, so that the system hyphenates the word if it does not fit properly at the end of the line. If the word does fit properly or, for example, more text is added which places the word on the following line anyway, the hyphen does not appear. See also *Hyphenation*.

DOWNLOADABLE FONTS

These are fonts that are bought separately and installed, to increase the variety of typefaces that are available on the printer. They may also be referred to as 'soft' fonts.

DTP

This stands for desktop publishing, which is the use of an electronic system, usually based on microcomputer technology, for the purpose of producing high-quality printed matter, which may include text, lines and graphic images.

EM DASH (—)

The size of this dash is the body width of the type size currently being used. It is used as a punctuation mark to break off one phrase from another where they do not flow naturally.

EN DASH (–)

An en dash is half the size of an em dash. See *Em Dash*.

EM SPACE

See *Fixed Spaces*.

EN SPACE

See *Fixed Spaces*.

EXPANDED TYPE

This is effectively the opposite to condensed type. For digital typesetters, altering the set width can cause the characters of a particular typeface to be expanded horizontally so as to occupy more width, for example, a 12-point character with a set width of 14 points.

>Normal type
>Expanded type

FIXED SPACES

In typography, where certain fixed spaces between characters are required, a system of fixed spacing based on points of space is employed to adjust the space between words or characters that is otherwise variable, depending upon the justification of the text on the line. There are three commonly used spacings:

em space — generally, space which is equivalent to the value of the point size, so that a 14-point em space will be 14 points wide;

en space — this is half an em space;

thin space — this may be either one quarter or one third of an em space.

These spaces are used to make adjustments where additional fixed space is required, for example the first line of most of the paragraphs in this book are indented from the left-hand margin.

FOLIO

The folio is the page number, usually placed at the top or bottom of a page, which may be aligned right, centred or aligned left according to the page layout. Folio is an old printing term for a page or sheet. See also *Footer* and *Header*.

FOOTER

This is usually a piece of information placed at the bottom of each page, though it may include a graphic, such as a line or logo, and it is often used to show such things as the page number, name of the publication, etc. It may also be called a trailer or running footer. See also *Header*.

FOUNT (FONT)

A complete set of characters in the same typeface and size, including letters, punctuation and symbols. For example, 12 pt Garamond is a different fount to 12 pt Garamond italic, or 12 pt Univers, etc.

FRAME

A non-printing border which defines the limits of a column of text or graphic image on the page. When displayed on the screen as faint dotted lines, frames are particularly useful when working with multiple-column documents, to see where the limits of text and graphic column boundaries are in relation to one another.

GALLEY

The word galley is generally only used in phototypesetting, since it refers to a continuous length of typeset material which may be used for proofing or cutting and pasting into position according to the layout requirements. Whereas DTP systems are designed to control the format and layout of the text before output, commercial typesetters would often provide the setting as lengths of text, set at the appropriate width. If this was then required for a manual, for example, the galley would need to be cut to the size of the pages, in order to achieve pagination and the text would then be pasted into its correct position on the pages.

The term galley originates from the 'hot metal' process of typesetting, where a metal tray with raised edges was used to hold about 20 inches of metal type. In this context galley referred to the amount of text set, but the word is more generally used to define the status of a job, i.e. a job which has reached 'galley stage'. For example, in book publishing it is usual for a book to be set in galleys, then these are proofed back to the author for checking. Once the final setting alterations are made, the galley is turned into paginated or formatted output according to the design requirements of the job.

GRAPHICS

Lines, rectangles, squares or circles created using the drawing tools of a desktop publishing system are graphics. They may also be bit-mapped images, either created using the DTP system or imported from another application or scanner. See also *Bit-mapping*.

GREEKING TEXT

When a page layout is reduced in size on the screen, often to facilitate the viewing of the full page area, the text may be represented as shaded blocks which exactly match the line breaks and position of the text. This may be necessary if the size of the text is too small at the reduced view to read legibly. Text displayed in this way is said to be 'greeked'.

GUTTERS

This refers to the space between columns in a multiple-column page layout, such as the gap between columns of a newspaper.

HAIRLINE

The thinnest possible rule (line), which is usually 0.25 of a point.

HALF-TONE

A printed illustration usually produced photographically, in which lights and shades of the original are represented by small and large dots, or varying concentrations of dots or lines.

HEADER

This is usually a piece of information placed at the top of each page, though it may include a graphic, such as a line or logo, and it is often used to show such things as the page number, name of the publication, chapter heading, etc. It may also be called a running header. See also *Footer*.

HYPHENATION

This is the process of deciding where to hyphenate the last word on a line. Some systems will do this automatically, based on a hyphenation dictionary, but you may wish to override this and include your own. See also *Discretionary Hyphen*.

ICON

A small graphic image on the screen that represents a function, object or pointer tool. For example, the image of a pencil to represent the tool for drawing.

INDENTS

Space defined at either end of the line to change the placement of the text. The most common form of indent is for a paragraph denoting the beginning of a block of text. Generally, you should make the indent space proportional to the length of the line, for example:

lines under 24 picas wide	indent 1 em space
lines between 25 and 36 picas wide	indent $\frac{1}{2}$ em spaces
lines 37 picas wide or over	indent 2 em spaces

JUSTIFIED TEXT

Text that is aligned on both the left- and right-hand margins of the document, as used in this book.

KERNING

Kerning is the reduction of space between characters of a typeface in order to improve the appearance when certain pairs of characters appear together. For example, the letter

W, when typeset, has a certain amount of optical space beneath the slant of the right and left downstrokes, and this can vary from typeface to typeface. If the letter following it, for example, is set with the normal letterspacing, it can appear to be displaced too far to the right. By subtracting points of space from between the characters, for example, the following letter effectively overlaps into the set width of the W, making the appropriate optical correction.

Paired kerning is the term used where a composition software system identifies the pairs of characters that require space adjustment and applies kerning automatically, provided that the facility is switched on. Generally, the kerning requirement is more noticeable the larger the size of the typeface.

Wa   Wa

Non-kerned Pair   *Kerned Pair*

LEADER

These are dots or dashes that fill the blank space left on a line and are usually inserted automatically by the DTP system when specified, for example:

    Chapter 1 About this Manual 15

LEADING

The amount of vertical space, expressed in points, between the baselines of two lines of text (it is pronounced 'ledding').

LETTER SPACING

This is the term that refers to the amount of space between individual letters, and on most phototypesetters is adjusted by subtracting or adding points of space. This is known as negative or positive letter spacing, respectively. Positive letter spacing may be used to put space between the letters of words, rather than have larger gaps between the words themselves which might make them look spread along the line width. Sometimes, positive letter spacing may be used to achieve a particular effect, for example, to space out the characters of a heading. Generally, positive letter spacing interrupts the legibility of the text, so negative letter spacing is the more commonly used adjustment. This is used to tighten the spacing between letters to improve the appearance of the text, adjust space for *kerning* or to fit copy into a tight space.

MACROS

A term used to describe a simple function (perhaps a single or combination of keystrokes) which performs a series of much more complicated tasks, which are usually user-definable.

MOUSE

This is a mouse-sized pointing device moved by hand on a desktop surface, which controls the position of a cursor or pointer on the screen in relation to the movements of the mouse itself. It also has up to three buttons on the top which can invoke various functions or selections. A mouse may either mechanically or optically translate its movements into co-ordinates for the screen cursor; optical mice require special surfaces on which to operate.

PAGE DESCRIPTION LANGUAGE (PDL)

This is a software facility which is independent of the hardware of desktop publishing systems. It is used to convert a screen image which may include text and graphics, into instructions that can be used to drive an output device like a laser printer. Using such an independent interface allows control over text and graphics, and multiple typestyles and sizes. PostScript is probably one of the best known page description languages, and is generally regarded as the current standard. It has been adopted by a wide range of manufacturers, including IBM and Apple, and offers a path to phototypesetting from PC front ends. PostScript was developed by Adobe. Other PDLs include Xerox Interpress, Imagen Impress and Imagen/HP DDL.

PAGE MAKE-UP SOFTWARE

This is software that can produce 'compound documents' which comprise text and graphics together. Using the concept of electronic paste-up, page make-up software can use text files generated on a word processor for pasting into a page layout. Graphics can be imported in a similar way or may be scanned in directly. Page make-up software normally uses a page description language. The distinction of such a package is that text-only software, such as a word processor, has limited output capabilities without a PDL, and provides limited control over layout and typefaces.

PAGINATION

Pagination is the process of assembling text, graphics, running headers, folios, etc., to produce pages of the required length ready for printing. This can be done by cutting and pasting-up setting and illustrations, etc., by hand, or electronically using a DTP or phototypesetting system.

PI CHARACTERS/FOUNTS

These are special characters or symbols such as $\frac{1}{4}$, $\frac{1}{2}$, $\frac{3}{4}$, $+$, $@$, $<$, $>$, etc. These may be used for mathematical requirements in text, or simply for decorative purposes.

PICA

A measurement equal to one-sixth of an inch.

POINT SYSTEM

A point is a unit of measurement generally considered to be equal to one seventy-second of an inch. However, there have been three point systems introduced:

- the American/British system: point is measured as 0.1383 inch, or one twelfth of a pica (pica being 0.166 inch);
- the Didot system: basic unit is the cicero, which is equal to 12 corps (points) or 0.178 inch. The Didot corps measures exactly 0.01483 inch;
- the Mediaan System: point (or corps) measures 0.01374 inch.

For general purposes, it is useful to remember: 6 picas to one inch, 72 points to one inch and 12 points to one pica. These, however, are not strictly accurate conversions since the point does not directly relate to inches.

PORTRAIT/LANDSCAPE FORMAT

Terms referring to the orientation of a page when printed. Portrait meaning the page is printed so that the longest sides are vertical (like the pages of this book), landscape is such that the sides of the page are the shortest.

PROPORTIONAL SPACING

Spacing between individual characters so that each character has an amount of horizontal space on the line proportional to the width of the character itself. For example, the letter 'i' will need much less space than the letter 'm'.

QUADDING

A term relating to the placement of text. The word originates from 'quadrat', which was a metal cube used for filling blank spaces in hand typesetting. Quads, therefore, are used to specify where the remaining space on a line is to be positioned.

This is quad left

                                                                                        This is quad right

                              This is quad centre

ROMAN

Ordinary letters, as distinct from italic, bold, etc. For example, this text is in roman. The term should not be confused with the typeface called Times Roman.

RULES

A term used in printing and design to refer to horizontal or vertical lines, for example as used in forms.

SERIF/SANS SERIF

Serifs are the short lines drawn at right angles to, or obliquely across, the ends of stems and arms of letters. The term is also used to describe typefaces that have this feature (e.g. Times). Sans serif is a typeface without serifs (e.g. Helvetica).

SET WIDTH

The set width of a character relates to its point size, so a 9 pt character has a set width of 9 points. Some typefaces, however, are designed with a narrower set width, e.g. 9 pt with 8.5-point set width.

SMALL CAPS

These are capital letters designed to match the x-height of a typeface, though many founts these days do not have small caps, so they are created by reducing the point size to 80% of its original size. They are used for abbreviations of awards, titles, etc., following a name, or where whole words in text are set in all caps.

CAPS and SMALL CAPS

SOFT FONTS

See *Downloadable Fonts*.

STYLE SHEET

All the styles that describe the design of a publication's layout and typefaces stored as a group, either in a file format or associated with a macro (see *Macro*).

SUPERIOR/INFERIOR CHARACTERS

These terms relate to characters usually set in a smaller size to the text typeface, and positioned above (superior) or below (inferior) the baseline. They are often referred to in DTP systems as superscripts and subscripts, respectively.

$$A^2 \quad B_4$$

TAGS

Labels given to paragraphs and other elements of a document which relate to a particular set of formatting instructions. For example, a paragraph tag may relate to the instructions that define the typeface, size, line length, spacing and other attributes.

TEMPLATE

Actually a style sheet, but often used as a model layout for similar publications. Some desktop publishing software packages provide a number of templates suitable for particular applications, such as, reports, magazines, pricelists, etc. (see *Style Sheet*).

TYPESIZE

Typesizes are described by the basic unit of measurement, the point. Points are used to define the length of a metal block or chunk of type. The length of the metal block's top surface relates to the point or typesize, though the character cast onto the surface, will be smaller than the overall size of the metal. Traditionally, then, the point size of a typeface refers to the dimension of the metal block, not the height of the image. This is to allow room for the ascenders and descenders. Thus the typesize relates to the distance between the top of the ascender and the bottom of the descender of the typeface.

TYPOGRAPHY

The design of printed matter. A typographer is one who is responsible for the character and appearance of printed matter. Once a specialist craft in itself, its importance has dwindled, sadly, with the emergence of desktop publishing as thousands of non-qualified 'amateur' typesetters control the appearance of all printed matter they originate, usually resulting is appalling design work.

WEIGHT

The weight of a typeface varies according to its design. The thickness of a line will determine how light or dark its image will appear after printing. There are standard terms for the various weights of a typeface and these are referred to as extra-light, light, semi-light, regular, medium, semi-bold, bold, extra-bold and ultra-bold. However, these variations are not standard for all typefaces. For example, 'medium' weight of one typeface may be the same weight as 'bold' in another.

WIDOWS AND ORPHANS

An orphan is the term used to describe the first line of a paragraph at the foot of a page when it is separated from the remainder which may be on the following page. A widow is the term used to describe the last line of a paragraph which appears at the top of a page, separated from the remainder which may be on the previous page.

WORD SPACING

In phototypesetting the space between words is variable (unlike a typewriter where the spacing is fixed), which enables lines of text to be justified. It also allows for the optimum number of words to be set on the specified line length.

WYSIWIG

(W)hat (Y)ou (S)ee (I)s (W)hat (Y)ou (G)et. This term is used to describe a feature of word processors or desktop publishing software systems that can represent on the screen display, what the actual output document will look like. The degree of accuracy in this respect, however, varies greatly from system to system.

X-HEIGHT

This is the height of the letter x and is used to describe the height of the body of lower case characters in a particular typeface. A change in x-height can affect the apparent size of typefaces that are actually the same size.

ZOOMING

The act of enlarging or reducing the view of a page layout. Enlarging the view of a portion of the document helps when working on detailed items, reducing the view allows the whole page or a pair of pages to be seen in full on the screen to check the overall appearance of the layout.

Index

A
abbreviations, 37
artificial intelligence, 114
author, 1–15
   hardware technical, 4–5
   management, 119–20
   recruitment, 119–20
   role of, 5–6
   software technical, 2–4
   training, 145–52

B
binding documentation, 66–9
   case binding, 69
   mechanical binding, 68
   perfect binding, 69
   ring binding, 68
   sewn binding, 69

C
CAD/CAM 101, 106, 107
capitals, 37
captions, 110
checklists for documentation projects, 19–21
City and Guilds, 145–6
clip art, 102–3
colour, use in technical documentation, 62
columnar text, 53–5
communication,
   skills, 12–14
   with others, 13–14

copyright, 37
costing, 125–9

D
database publishing, 98–9
dates, 37
design of documentation, 51–73
desktop publishing, 97–9
   graphics, 91
   output devices, 92
   style control, 89
   typefaces, 90–1
   see also electronic publishing
desktop publishing systems,
   categories of, 92–8
   dedicated electronic publishing systems, 92, 97–8
   PC-based electronic publishing, 92, 96–7
   PC-based page make-up, 92, 94–6
   word processor based, 92, 93
development of documentation, 136–7
distance learning, 152–3
   see also training of authors, 147
documentation types, 42–8
drawing software, 102
DTP, *see* desktop publishing

E
editing, 26–7
education, *see* training of authors

electronic publishing, 87–99
  see also desktop publishing
expert system, 114–15

F
figures, 37
  see also illustrations
finishing of documentation, 63–72
footers, 55
formatting, 21–2

G
generic coding, 96
glossary of terms, 177–88
grammar, 12–13
  checkers, electronic, 78, 79–83
graphics, 101–11
  including in documents, 106–8
  positioning, 108–9
  software, 102
  see also illustrations

H
headers, 55
headings, 36
help systems, 113–14
  see also on-line documentation
higher education, 147
  see also training of authors, 147
house styles, 36–8
hypertext systems, 114–16
  see also on-line documentation
hyphenation, 36

I
illustrations, 55, 101–11
  captions, 110
  positioning, 108–9
  referencing, 109
  see also graphics
image file formats, 104–6
  BMP, 104–5
  DIB, 104–5
  GIF, 105
  IMG, 106
  MAC, 106
  PCX, 105
  RLE, 104–5
  TGA, 106
  TIFF, 105–6
  Vector, 106

imposition, 69–71
in-house publication systems, 124–5
indexing, 38
  automatic, 84
institutions, 157–69
international documentation, 141–4

L
language, 23–6
laser printing, 92
layout, 52–8
lists, 37

M
maps, 37
multimedia systems, 114–16
  see also on-line documentation

N
notes, 37

O
objectives for documentation, 18–19
on-line documentation, 113–16
organisational skills, 14–15
organisations, 157–69
overhaul documentation, 45

P
packaging, 63–72
page layout, 53
page make-up, 89
page size, 52–3
page style, 36
paper sizes, 64–6
paper types, 71–2
  art paper, 72
  cartridge, 72
  wood-free, 72
photographs, 107
phototypesetting, 92
planning documentation jobs, 17–21, 118–19
plates, 37
pre-press preparation, 38–9
print buying, 120–4
print specifications, 121–3
printers, types of, 92
proof correction marks, 28, 29–35
proof-reading, 27–8
publications management, 117–40

publications manager, role of, 117–40
punctuation, 36

Q

qualifications, 145–7
   see also training of authors
quality control, 131–3

R

readability indices, 81–3
   Bormuth grade level, 83
   Coleman–Liau grade level, 82
   Flesch reading ease index, 82
   Flesch–Kincaid grade level, 82
reader knowledge, 9–12
recruitment of authors, 119–20
reference manuals, 44
reference system, 37
referencing, automatic, 84
reports, 46
revisions of documentation, 47–8, 118–19

S

scanning graphic images, 107
scheduling, 125, 129–31
service documentation, 45
services for technical documentation, 162–9
spacing, 62
specifications, 28
spelling, 36
   checkers, electronic, 78–9
standards, 155–7
style, 23–6
   checkers, electronic, 78, 79–83
subcontracting documentation, 133–6
subject knowledge, 7
supplier selection, printers, 123–4
support documentation, 44–5

T

tables, 37
technical author, 1–15
   hardware, 4–5
   software, 2–4
   see also author
technical documentation services, 162–9
technical publications, 41–9
technical writing, 17–40
template, document, 94
text highlights, 37
thesauruses, electronic, 79–83
trademarks, 37
training material, 46–7
training of authors, 145–52
translation, 143
   see also international documentation
trimming of finished documents, 66
troubleshooting guides, 46
typefaces, 59–60
typesize, 60
typestyles, 60
typography, 59

U

units, 37
upgrades to documentation, 47–8
user guides, 42–4

W

word processing, 75–85
word processing features, 75–8
   alignment, 76–7
   editing functions, 76–7
   formatting, 76–8
   spacing, 76–7
writing sources, 7–9